This Belongs To:

Munroe-Meyer Institute for Genetics & Rehabilitation
Physical Therapy Department
Occupational Therapy Department
985450 UNMC
Omaha, Nebraska 68198-5450
(402) 559-6415 or 1-800-656-3937

Purchase date 6-03 Cost 65⁰⁰ Copy 1 of 1

Progress in Motor Control

VOLUME TWO

Structure-Function Relations in Voluntary Movements

Mark L. Latash, PhD
The Pennsylvania State University

Editor

Human Kinetics

Library of Congress Cataloging-in-Publication Data
Progress in motor control / Mark L. Latash, editor.
    p.   cm.
    Includes bibliographical references and index.
    Contents: v. 2. Structure-function relations in voluntary movements.
    ISBN 0-7360-0027-5
    1. Human locomotion.   2. Motor ability.   3. Bernshteîn, N.A.
    (Nikolaî Aleksandrovich), 1896-1966.   I. Latash, Mark L., 1953
    QP301.P767   1998
    612.7'6--DC21   97-44950
                                                                              CIP

ISBN: 0-7360-0027-5

Acquisitions Editor: Judy Patterson Wright, PhD; **Managing Editor:** Amy Stahl; **Assistant Editor:** Derek Campbell; **Copyeditor:** Barbara J. Field; **Proofreader:** Erin Cler; **Indexer:** Marie Rizzo; **Graphic Designer:** Fred Starbird; **Graphic Artist:** Kathleen Boudreau-Fuoss; **Photo Manager:** Leslie A. Woodrum; **Cover Designer:** Fred Starbird; **Cover photo:** Courtesy of *Fizkultura i Sport*; **Art Manager:** Carl D. Johnson; **Printer:** Edwards

Printed in the United States of America   10   9   8   7   6   5   4   3   2   1

Human Kinetics
Web site: www.HumanKinetics.com
United States: Human Kinetics, P.O. Box 5076, Champaign, IL 61825-5076
800-747-4457
e-mail: humank@hkusa.com
Canada: Human Kinetics, 475 Devonshire Road Unit 100, Windsor, ON N8Y 2L5
800-465-7301 (in Canada only)
e-mail: orders@hkcanada.com
Europe: Human Kinetics, 107 Bradford Road, Stanningley, Leeds
LS28 6AT, United Kingdom
+44 (0) 113 255 5665
e-mail: hk@hkeurope.com
Australia: Human Kinetics, 57A Price Avenue, Lower Mitcham, South Australia 5062
08 8277 1555
e-mail: liahka@senet.com.au
New Zealand: Human Kinetics, P.O. Box 105-231, Auckland Central
09-523-3462
e-mail: hkp@ihug.co.nz

Contents

Preface

The second volume of the series *Progress in Motor Control* continues a tradition started in 1998 when the first volume, *Bernstein's Traditions in Movement Studies,* was published by Human Kinetics. Similar to the first volume, the present book contains chapters written by the invited speakers who took part in the Second International Conference "Progress in Motor Control: Structure-Function Relations in Voluntary Movement."

Although the subtitle of this volume does not mention Nikolai Bernstein, the spirit of this great scientist could definitely be perceived during the conference and, I hope, is also present in the chapters of this book. Bernstein contributed significantly to the structure-function controversy and wrote two articles in which he expressed and developed his rather revolutionary views on the relationship between structure and function in the central nervous system. The first article was written in 1935. Its English translation was published in 1967, and it has enjoyed considerable international attention. The second article was written by Bernstein and his two younger colleagues, Philip Bassin and Lev Latash (Bassin et al., 1965). It was published in 1965 in a rather obscure book and, for a long time, remained known only to a handful of Russian colleagues. An English translation of this article has recently been published in *Motor Control* within its section "Bernstein's Heritage" (Latash et al., 1999, 2000). Many of the ideas expressed in the latter article are far from obsolete, and Bernstein's contribution to the structure-function discussion is far from completed.

The first book of the series had many features that are preserved in the second volume. In particular, several chapters were written by scientists from the former Soviet Union who either knew Bernstein personally or worked closely with Bernstein's students. These include Yuri Arshavsky, Tatiana Deliagina, Israel Gelfand, Marat Ioffe, and Grigory Orlovsky.

The current volume is written with the assumption that readers have a considerable background in movement studies. It is a reference book for professionals working in motor control and in related areas such as biomechanics, neurophysiology, psychology, movement disorders and rehabilitation, and motor development. The book can also be used as additional material by advanced undergraduate and graduate students in these areas. Each chapter contains an introduction and a conclusion; these elements guide readers through the main body of the chapter.

The first volume of *Progress in Motor Control* focused on theories and hypotheses in the field. Frequently, hypotheses or models are developed and described in formal language that relates only metaphorically to the neurophysiology and biomechanics of the human or animal body. For many potential readers, such general accounts may be of limited use because they do not provide links to the structures in the human body in which hypothesized interactions can take place. This limitation is particularly true for those readers who study the control of voluntary movements in applied areas such as movement disorders, motor rehabilitation, effects of practice on movements, and changes in movements during development and aging. Hence, a main goal of the current volume is to provide support for or criticism of some of the current theories and hypotheses using data obtained in human and animal experiments as well as in clinical studies.

The chapters in this volume follow the traditions set by Bernstein in his seminal works and suggest direct relations between aspects of the motor function and biomechanical or neurophysiological structures. This volume opens with two chapters written by two of the most productive researchers, Apostolos Georgopoulos and John Rothwell, who perform neurophysiological studies of the brain in relation to voluntary and involuntary movements. These chapters review recent findings in the respective research groups that link the activity of cortical neurons and of neurons in the reticular formation to the generation of voluntary movements.

Chapters 3, 4, and 5 address the issue of postural control using a wide range of objects and methods of study. Simon Bouisset and Serge Le Bozec review the vast experience of their group in biomechanical studies of postural preparation to predictable perturbations. A Russian team led by Grigory Orlovsky shows that neurophysiological studies of relatively simple animals allow deep insights into the organization of systems responsible for postural stabilization. Marjorie Woollacott reviews her unique experience with studies of motor development, particularly with respect to postural control, in typically developing children and in those with cerebral palsy. Dr. Woollacott addresses an important issue for those who deal with motor behavior in atypical populations: What is the relative role of peripheral (musculoskeletal) and central (neural) factors in the apparently suboptimal motor patterns seen in such children?

In chapter 6, the clinical theme is continued by Mindy Levin and her coauthors, who study joint coordination during simple reaching movements in two groups of patients: those with hemiparesis after stroke and those with cerebral palsy. This chapter provides convincing evidence for atypical patterns of joint coordination in such patients. Another applied aspect of movement studies is addressed in chapter 7 by M. Ioffe and his colleagues using neurophysiological studies of the relations between the activity in particular cortical areas and changes in motor patterns with practice.

Chapters 8, 9, and 10 focus on the role of biomechanical factors in the production of movements. A new model based on a fractional power muscle damping is presented by Jim Houk and his colleagues. Richard Nichols and his colleagues review an impressive series of studies showing how spinal reflex loops can be used to modulate the mechanical properties of a limb to assure its stability in various motor tasks. In chapter 10, Roger Enoka and his coauthors review their experimental material related to the specificity of lengthening as opposed to shortening muscle contractions.

The last two chapters return to more general issues. Chapter 11, which revisits and develops the ideas of Nikolai Bernstein, is coauthored by Israel Gelfand, who wrote seminal papers on theoretical issues of motor coordination in the early 1960s.

The chapter reviews those studies and moves ahead in the quest for a set of adequate notions for analysis of biological movement and possibly of other biological phenomena. Onno Meijer is well known to the motor control community for his brilliantly controversial papers on the history and philosophy of movement studies. In the final chapter of this volume, Meijer reviews one of the most fascinating and controversial disagreements—between Ivan Pavlov and Nikolai Bernstein—in the history of contemporary neurophysiology.

Publication of this book would have been impossible without the generous financial support of the National Science Foundation, Whitaker Foundation, Continuing and Distance Education, the Department of Kinesiology of the Pennsylvania State University, and Human Kinetics Publishers. I would also like to thank personally those who shared with me the load of organizing and running the conference: Katherine Deutsch, Karl Newell, David Rosenbaum, Dagmar Sternad, and Vladimir Zatsiorsky.

Mark L. Latash

References

Bassin, Ph.V., Bernstein, N.A., and Latash, L.P. (1966) On the problem of relation between brain structure and function in its contemporary understanding. In Grastchenkov, N.I. (Ed.), *Physiology in clinical practice,* pp. 38-71. Moscow: Nauka.

Bernstein, N.A. (1935) The problem of interrelation between coordination and localization. *Archives of Biological Sciences,* 38: 1-35.

Bernstein, N.A. (1967) *The co-ordination and regulation of movements.* Oxford: Pergamon Press.

Latash, L.P., Latash, M.L, and Meijer, O.G. (1999) Thirty years later: On the problem of the relation between structure and function in the brain from a contemporary viewpoint (1966). Part I. *Motor Control,* 3: 329-345.

Latash L.P., Latash M.L, and Meijer, O.G. (2000) Thirty years later: On the problem of the relation between structure and function in the brain from a contemporary viewpoint (1966). Part II. *Motor Control,* 4: 125-149.

Credits

Figure 3.1—Reprinted, by permission, from M.P. Murray, A. Seireg, and R.C. Scholz, 1967, "Center of gravity, center of pressure and supportive forces during human activities," *Journal of Applied Physiology* 23: 831-888.

Figure 3.3—Reprinted, by permission, from *Journal of Biomechanics*, 20(8), S. Bouisset and M. Zattara, "Biomechanical study of the programming of anticipatory postural adjustments associated with voluntary movement," 735-742, Copyright 1987, with permission from Elsevier Science.

Figure 3.5—Reprinted, by permission, from S. Bouisset, J. Richardson, and M. Zattara, 2000, "Do APAs occurring in different segments of the postural chain follow the same organisation rule for different task movement velocities, independently of the inertial load value?" *Exp Brain Res* 132: 83. Copyright © Springer-Verlag

Figure 3.7—From M.C. Do, S. Bouisset, and C. Moynot, 1985, "Are paraplegics handicapped in the execution of a manual task?" *Ergonomics* 28(9):1363-1375.

Figure 4.03, a-d—Reprinted, by permission, from T.G. Deliagina, G.N. Orlovsky, A.I. Selverston, et al., 1999, "Neuronal mechanisms for the control of body orientation in *Clione* I. Spatial zones of activity of different neuron groups," *Journal of Neurophysiology,* 82: 687-699.

Figure 4.04—Reprinted with permission from *Nature* 393: 172-175, "Control of spatial orientation in a mollusc," by T.G. Deliagina, Y.I. Arshavsky, and G.N. Orlovsky. Copyright 1998 Macmillan Magazines Limited.

Figure 4.06, a-f—Reprinted, by permission, from T.G. Deliagina, G.N. Orlovsky, A.I. Selverston, et al., 1999, "Neuronal mechanisms for the control of body orientation in *Clione* I. Spatial zones of activity of different neuron groups," *Journal of Neurophsiology,* 82: 687-699.

Figure 4.07, a and b—Reprinted with permission from *Nature* 393: 172-175, "Control of spatial orientation in a mollusc," by T.G. Deliagina, Y.I. Arshavsky, and G.N. Orlovsky. Copyright 1998 Macmillan Magazines Limited.

Figure 4.08, a-f—Reprinted with permission from *Nature* 393: 172-175, "Control of spatial orientation in a mollusc," by T.G. Deliagina, Y.I. Arshavsky, and G.N. Orlovsky. Copyright 1998 Macmillan Magazines Limited.

Figure 4.10c—Reprinted by permission from *Experimental Brain Research,* "Vestibular control of swimming in lamprey. II," T.G. Deliagina, G.N. Orlovsky, S. Grillner, et al., 90: 489-498, Figure 2.a, 1992, Copyright © Springer-Verlag.

Figure 4.10d—Reprinted by permission from *Experimental Brain Research,* "Visual input affects the response to roll in reticulospinal neurons of the lamprey," T.G. Deliagina, G.N. Orlovsky, S. Grillner, et al., 95: 421-428, Figure 4.a, 1993, Copyright © Springer-Verlag.

Figure 4.13, a and b—Reprinted, by permission, from T.G. Deliagina and P. Fagerstedt, 2000, "Responses of reticulospinal neurons in intact lamprey to vestibular and visual inputs," *Journal of Neurophysiology* 83: 864-878.

Figure 4.15, a-d—Reprinted, by permission, from T.G. Deliagina, P.V. Zelenin, P. Fagerstedt, et al., 2000, "Activity of reticulospinal neurons during locomotion," *Journal of Neurophysiology* 83: 853-863.

Figure 4.16, a-f—Reprinted, by permission, from P.V. Zelenin, T.G. Deliagina, S. Grillner, et al., 2000, "Postural control in the lamprey: A study with neuro-mechanical model," *Journal of Neurophysiology* 84: 2880-2887.

Figure 5.1—Reprinted, by permission, from M.N.C. Roncesvalles, M.H. Woollacott, and J.L. Jensen, 2001, "Development of lower extremity kinetics for balance control in infants and children," *Journal of Motor Behavior* 33 (2): 180-192. Reprinted with permission of the Helen Dwight Reid Educational Foundation. Published by Heldref Publications, 1319 18th Street, NW, Washington, DC 20036-1802. Copyright © 2001.

Figure 5.2—Reprinted, by permission, from H. Sveistrup and M.H. Woollacott, 1996, "Longitudinal development of the automatic postural response in infants," *Journal of Motor Behavior* 28 (1): 58-70. Reprinted with permission of the Helen Dwight Reid Educational Foundation. Published by Heldref Publications, 1319 18th Street, NW, Washington, DC 20036-1802. Copyright © 2001.

Behavioral and Neural Aspects of Motor Topology

Following Bernstein's Thread

Apostolos P. Georgopoulos

Brain Sciences Center,
Minneapolis Veterans Affairs Medical Center;
Departments of Neuroscience, Neurology, and Psychiatry,
University of Minnesota Medical School;
and Cognitive Sciences Center,
University of Minnesota

In his article "The Problem of the Interrelation of Co-ordination and Lateralization," Bernstein (1935) drew attention to invariances in the shape of drawings made under very different conditions, such as using different effectors or different combinations of muscles and joints. He called these invariances "topological" and contrasted them with other "metric" aspects of movement, such as size and location in space. He then speculated on the brain representation of motor topology, as follows:

> *"There is the deeply seated inherent indifference of the motor control centre to the scale and position of the movement effected. . . . It is clear that each of the variations of a movement (for example, drawing a circle large or small . . .) demands a quite different muscular formula; and even more than*

this, involves a completely different set of muscles in the action. The almost equal facility and accuracy with which all these variations can be performed is evidence for the fact that they are ultimately determined by one and the same higher directional engram in relation to which dimensions and position play a secondary role." (Bernstein, 1935; see Whiting, 1984, p. 109)

This insight of Bernstein's, from 45 years ago, into the "directional engram" is remarkable. In a way, our work during the past 20 years on the neural coding of motor direction can be seen as addressing precisely this issue (i.e., the extraction of directional information from the impulse activity of single cells and cell populations in cortical areas). The discovery of directional tuning provided the key link between neural activity and the direction of movement and made possible the neural construction of a motor trajectory in space (Georgopoulos et al., 1988; Schwartz, 1994). This "neural trajectory" proved to be an accurate and isomorphic representation of the actual motor trajectory. Remarkably, this was also predicted by Bernstein (1935), who stated that "the higher engram . . . is extremely geometrical, representing a very abstract motor image of space" (Whiting, 1984, p.109). Indeed, space is pervasive in figure drawing. Unlike relatively pure temporal functions, such as tapping, figure drawing cannot be conceived apart from the spatial relations connecting the elements of the figure. Therefore, the geometric aspects of the shape are of fundamental importance for its drawing.

In this chapter, I review the results of recent studies that have dealt with issues at the heart of Bernstein's concerns: motor topology and the neural representation of direction, size, and location of movement in space. Specifically, I discuss the results of behavioral studies of drawing geometrical figures and then turn to the issue of representation of topological features and how these can be invariantly extracted from neuronal populations irrespective of attributes of size, location, and the muscles effecting the motor trajectory. Finally, I discuss applications of such neurally inspired operations to artificial neural networks trained to draw figures.

Drawing Figures: Bernstein's Perspective

A major theme in Bernstein's thought was "motor topology." By that he meant the *relative* spatial features of a drawn figure—relative, that is, among its parts and irrespective of the size and location of the figure or the dynamic aspects of the movement, such as speed of drawing. These invariances have profound implications for the motor system, for they correspond to drastically different configurations of joints and patterns of muscle activations. Thus, a square can be drawn large or small, in different parts of space, at different speeds, in movement or isometric force space, and in different sequences of drawing its sides, while always retaining the proper relative proportions among its parts that make the figure a square. Although the *execution* of motor commands is ultimately specified in terms of torques or muscle length-tension curves (Hinton, 1984), the *representation* and *planning* of the relative spatial aspects of the motor command in terms of different joint and/or muscle configurations, and even different effectors, make figural invariances a very special case. First, I deal with a specific aspect of Bernstein's motor topology: the classification of figures according to their spatial features.

Motor Topology

Figures can be classified according to different sets of features. In discussing this topic, Bernstein wrote: "As topological properties of a linear figure, for example, we may discuss whether it is open or closed, whether the lines composing it intersect with each other as in a figure eight or whether they do not intersect as in the case of a circle and so on" (Bernstein, 1935; see Whiting, 1984, p. 103). The key outcome of any classification scheme is grouping figures (and objects in general) together according to their common features. Now, there may be a multitude of such features (e.g., perimeter, number of sides, orientation, openness), some of which may covary with others such that classification schemes rarely rely on a single feature. Moreover, this discussion is obviously most clearly applicable to percepts of figures rather than their motor drawings. Therefore, another approach is needed beyond the parametric, feature-based procedure. Indeed, a very general, nonparametric, all-encompassing concept in this regard is that of *similarity*. Apparently, objects are commonly grouped according to how similar they are, and although groups of dissimilar objects might be distinguished on the basis of differing distinct features, it is usually combinations of features, rather than single features, that underlie dissimilarity judgments. However, the important point is that, as Shepard (1980) pointed out, "Without any quantitative information about the physical properties of colors, tones, speech sounds, or words, we can learn something about how humans process such stimuli from an analysis of ratings of perceived similarity" (p. 390).

Let me now turn to the application of these ideas to motor topology. First, I want to draw attention to the possibility that the same figures might be grouped differently depending on whether one refers to their percepts or motor drawings. This is because the same features (e.g., perimeter, number of sides) might have a different impact on the perception or the motor drawing of a figure. My colleagues and I (Averbeck et al., 1998) investigated this problem in a recent study in which we analyzed perceptual and motor dissimilarities among nine geometric figures of approximately the same surface area. In the motor experiment, subjects were shown a figure and asked to copy it using a handheld joystick; in the perceptual experiment, subjects were shown pairs of figures and were asked to indicate in an analog scale the degree of perceived dissimilarity between the two figures by positioning a pointer along a line. The perceptual experiment yielded direct pairwise dissimilarity judgments; motor dissimilarities were derived from the figure drawings by calculating the area of nonoverlap between pairs of figures.

There are essentially two ways to analyze dissimilarity data: multidimensional scaling (MDS) and tree modeling (Shepard, 1980). These analyses address different questions. Specifically, MDS aims at reducing a high-dimensional object space to, commonly, a two-dimensional (2-D) space while keeping the interobject distances in the reduced space similar to those in the high-dimensional space; the outcome of the analysis is a 2-D "object configuration" plot in which the objects are plotted at the coordinates derived by the analysis. In this plot, relative relations among objects can be visualized more easily, and if the new dimensions are readily interpretable, a meaningful remapping of the objects in the new 2-D space can be obtained. On the other hand, tree modeling aims at grouping the objects according to common properties. This analysis yields a tree with branches containing similar objects, but what it is that groups objects together is usually a challenge in interpreting the plots, much as the interpretation of dimensions is for MDS. We applied both of these analyses on the perceptual and motor dissimilarity data obtained in the experiment mentioned earlier. We found that perceptual judgments on, and motor drawings of, the nine figures used differed in their MDS object configuration space plot and in their additive tree modeling plot. This finding indicates that brain systems for perceptual analyses and motor drawings of figures rely on different aspects of the figures.

The methods of MDS and tree modeling provide useful tools with which to address experimentally the issue of what Bernstein called "topological class" (Bernstein, 1935; see Whiting, 1984). In his extensive discussion of this issue (Bernstein, 1935; see Whiting, 1984), Bernstein enumerate several features of objects, which he calls "topological properties" and which bear on classifying objects in different groups. He exemplified his train of thought by pointing out that "every printed letter is a separate topological class of the first order, while to the single class of letter *A* there belong letter *A*'s of all dimensions, scripts, outlines, embellishments, etc." (Bernstein, 1935; see Whiting, 1984, p. 104). Obviously, Bernstein's approach for classification is a mixture of explicit feature description and symbolic considerations; for example, the class of letter *A* is partly a symbolic one, for some *A*'s could conceivably be readily confused with another letter. This means that the allocation of a given object to a specific class might be ambiguous depending on the formal criteria used or the actual placement of the object in a specific class by a particular subject, irrespective of those formal criteria. It is precisely here that the assessment of dissimilarity by experimental methods becomes important; in a sense, the procedure is reversed: instead of classifying objects based on theoretical considerations of features, classification schemes are derived from experimentally assessed dissimilarities, and these schemes are then interpreted based on object features used by the brain. Now, since features may be fuzzy at times and frequently not entirely independent of each other, it is reasonable to suppose that dissimilarities may be based on more "holistic" aspects of the objects. Indeed, this state of affairs captures much better Bernstein's original intuition when he stated that "by the topology of a geometrical object I mean the totality of its qualitative peculiarities" (Bernstein, 1935; see Whiting, 1984, p. 103). It is this elusive totality on which the vague but powerful concept of (dis)similarity relies.

Effects of Brain Damage on Figure Copying

The translation of a seen figure to a drawing is accomplished by appropriate movements of the hand, produced, in turn, by the generation of suitable torques at various joints. This function of copying is frequently disturbed in patients with cortical damage who may be impaired in copying even simple geometrical figures such as a square. In some cases, copying is disturbed in the presence of normal object recognition (e.g., Cipolotti and Denes, 1989), whereas in other cases, figure drawing is normal in the presence of severe object agnosia (e.g., Behrmann et al., 1992). Finally, an additional dissociation concerns drawing from vision versus drawing from memory: either of the two can be intact while the other is disturbed (e.g., Servos and Goodale, 1995). All of this evidence indicates that the neural mechanisms underlying spatially patterned drawing are separate from those subserving perceptual recognition. Moreover, the disorder is not specific to drawing, for it is also manifested in assembly tasks, such as when a patient is asked to form a square by suitably arranging four matchsticks. This indicates that the problem is with the relative spatial (topological) relations of figural elements irrespective of the kind of motions (or effectors) used. This idea is also supported by the fact that, even in drawing, a square can be formed by drawing the sides in different sequences; in fact, a good-looking square was drawn discontinuously (i.e., by drawing nonadjacent sides) by a patient suffering from severe object agnosia (Behrmann et al., 1992). The neural mechanisms underlying these intrafigure spatial relations are essentially unknown. Although early studies pointed to a special role of the right cerebral hemisphere in constructional apraxia (Benton, 1967; Mack and Levine, 1981; Piercy et al., 1960), a disorder characterized by impairment in copying, more systematic later work (reviewed in De Renzi, 1982; Gainotti, 1985) suggested that both hemispheres are probably involved in this function.

Neural Coding of Figure Elements

In contrast to the paucity of knowledge concerning the brain mechanisms of motor figures as wholes, the neural coding mechanisms underlying the drawing of the elements of a line figure (i.e., the drawing of simple lines) have been well investigated and partially elucidated. The approach has been to train monkeys to draw lines in different directions and amplitudes and then record the activity of single cells in the brain during drawing. Such studies have provided the following information (see Georgopoulos, 1995, for a review). First, the activity of single cells in several motor structures is tuned with respect to the direction of movement in space. Specifically, there is a particular direction of movement for which a cell would discharge at the highest rate, called the cell's *preferred direction*. Second, cell activity is also modulated by the amplitude of the movement (Fu et al., 1993). The directional tuning is observed during both the reaction time and the movement time, whereas the modulation with movement amplitude is observed mostly during the movement time (Fu et al., 1995). Third, the preferred direction of a cell is usually stable (e.g., it is invariant for different movement amplitudes (Fu et al., 1993), but it can be influenced by the posture of the arm, such as when the initial position changes (Caminiti et al., 1990) or when the posture is explicitly altered by a special mechanical arrangement (Scott and Kalaska, 1997). These findings indicate an orderly relation between cell activity and spatial movement parameters. However, the directional tuning by itself does not provide for a unique coding of the direction of movement at the single cell level since the tuning is broad (but not as broad as previously thought [Amirikian and Georgopoulos, 2000]) and the preferred direction can be affected by posture. Any unique information, then, should rely on the neuronal ensemble of directionally tuned cells. Indeed, this information can be extracted by various kinds of population vector codes (Georgopoulos et al., 1983, 1986; Salinas and Abbott, 1994). The outcome of these decoding schemes is the vector sum of weighted vectorial contributions of individual cells, called the *neuronal population vector* (NPV) (Georgopoulos et al., 1983, 1986, 1988), which has the following useful characteristics. First, the direction of the NPV is close to the direction of movement in space. Thus, the population decoding transforms aggregates of purely temporal spike trains into a directional signal, isomorphic to the direction of movement. Second, the calculation of the NPV is a rather simple procedure, for it (a) rests on the directional selectivity of single cells, which is apparent; (b) involves weighting of vectorial contributions by single cells on the basis of the change in cell activity, which is reasonable; and (c) relies on the vectorial summation of these contributions, which is practically the simplest procedure to obtain a unique outcome. In fact, an important aspect of the population vector analysis is that it relies on the directional tuning as defined operationally by the procedures mentioned earlier; no special assumptions are made or required as to how this tuning comes about. Finally, the NPV is robust. It is a distributed code and as such does not depend exclusively on any particular cell. Its robustness is evidenced by the fact that it can convey a good directional signal with only a small number of cells (Georgopoulos et al., 1988; Salinas and Abbott, 1994). However, a much more important property of the NPV is that it is an unbiased predictor of the direction of movement when the posture of the arm changes (Caminiti et al., 1990; Kettner et al., 1988; see also figure 1 in Georgopoulos, 1995), even when the preferred directions of individual cells may change with different movement origins (Caminiti et al., 1990). The NPV then provides posture-free information about the direction of movement in space.

This spatial invariance of the NPV shines when it is used to construct "neural trajectories" of continuous tracing movements (Schwartz, 1994). In these experiments, monkeys are trained to track a moving light along a predetermined trajectory, such as a spiral, circle, ellipse, or figure eight. The NPV is calculated at short time intervals (e.g., 10-20 ms) and strung tip-to-tail to form a neural trajectory. This neural trajectory has consistently been found to be an excellent predictor of the actual trajectory. The reason for this is that not only does the direction of the instantaneous NPV predict the direction of the movement but also its intensity predicts the instantaneous amplitude of the movement (i.e., speed for a fixed time bin) such that the spatially arranged NPV time series yields the upcoming movement trajectory in space.

All of these findings show that accurate and robust information about movement direction can be extracted from an ensemble of directionally tuned neurons in the form of the NPV, which is invariant with respect to the posture of the arm and which, in drawing movements, carries information about the whole trajectory.

Figure Drawing by Artificial Neural Networks

The time-varying directional operations discussed previously have been modeled using a massively interconnected artificial neural network that consists of directionally tuned neurons and produces as an outcome the neuronal population vector (Lukashin and Georgopoulos, 1993, 1994a, 1994b). This network has been successfully trained to draw accurate geometric shapes (Lukashin et al., 1994, 1995) and has led to a novel hypothesis on how memorized trajectories of complex movements could be stored in the synaptic connections of overlapping neural networks (Lukashin et al., 1994). The idea is that there is a general-purpose network that is involved in all kinds of movements, memorized or not, but which carries no information about memorized trajectories of specific shapes (e.g., circles, ellipses, scribbles) and which, if activated alone, would produce straight line trajectories. It is now hypothesized that there are also networks highly specific to a particular trajectory (e.g., clockwise circle) that are interconnected with the general-purpose network; when a specific trajectory needs to be performed, the appropriate specific network fuses with the general-purpose network and, now as one network, produces the desired trajectory. Remarkably, the size of the specific network need only be less than 5% of the size of the general-purpose network for the desired trajectory to be effectively stored and reproduced (Lukashin et al., 1994). It is noteworthy that such very specific cells have been observed at low proportions in neurophysiological recordings during performance of memorized trajectories (Ashe et al., 1993; Hocherman and Wise, 1991).

The idea of the existence of very specialized networks raises the question of the degree of specialization and of how such networks are created in the first place. We can only speculate on these issues. With respect to the general-purpose network, it is reasonable to assume that it is present at birth since it is assumed to subserve all movements. There are several possibilities concerning the specialized networks. One possibility is that a number of small-size networks, specific for basic shapes (e.g., straight lines, curves, and some combinations thereof ["motor shape primitives"]), are present at birth. Then motor learning for other, complex motor acts would consist of adjusting the connection strengths between the general-purpose and specific networks. This idea implies that all of the specialized primitives are used routinely, although not as frequently as the general-purpose network. Another possibility is that innate specific networks code for more complicated shapes and are large in number. The mechanism of motor learning would then be similar to that described previously; but in this case, only a small number of the specialized networks would be used. This means that a number of the complex specialized networks may never be used. This situation would be similar to that encountered in the immune system, in which there is a large potential for making a large number of antibodies, of which, however, only some may actually be made, depending on the exposure of the organism to specific antigens. In both cases there is a selection: selection of a specialized trajectory or selection of an antibody, both from a large ensemble available. Finally, an intermediate hypothesis would be that we start with motor shape primitives, but that the more complex trajectories resulting from the combination of these primitives with the general network become themselves very specialized and behave as such in the formation of other trajectories in novel associations.

Neural Representation of Memorized Figures

The neural mechanisms subserving well-learned, memorized, complex movement trajectories are elusive. This problem was investigated in a study (Ashe et al., 1993) in which monkeys were trained to perform from memory an arm movement with an orthogonal bend, up and to the left, following a waiting period. They held a 2-D manipulandum over a spot of light at the center of a planar working surface. When this light went off, the animals were required to hold the manipulandum there for 600-700 ms and then move the handle up and to the left to receive a liquid reward. There were no external signals concerning the go time or the trajectory of the movement. Following 20 trials of the memorized movement trajectory, 40 trials of visually triggered movements in radially arranged directions were performed. The activity of 137 single cells in the motor cortex was recorded during performance of the task. A high percentage (62.8%) of cells changed activity during the waiting period. Other cells did not change activity until after the 600-ms minimum waiting time was over, and occasionally cell activity changed almost exactly 600 ms after the center light was turned off. However, the most interesting observation was that a few cells changed activity *exclusively during the execution of the memorized movement* (see figure 5 in Ashe et al., 1993); these cells were completely inactive during performance of similar movements in the visually guided control task. These findings suggest that *performance of a movement trajectory from memory may involve a specific set of cells,* in addition to the cells activated during both visually guided and memorized movements. This idea is in accord with the results of the modeling studies summarized earlier.

Coding of Dynamic Isometric Force

Although the exertion of static force is not uncommon, as when we hold a book against gravity, the production of a change in force is almost universally present in all actions, for it precedes the beginning of the movement. Study of force change under conditions of limb motion is hampered by several factors that cannot be easily controlled experimentally (e.g., interaction forces). An experimentally "clean" case is provided by the exertion of an isometric force pulse (i.e., by the production of a rapid change in force in the absence of limb motion). Such a paradigm, which allows the study of the relation of neural activity to pure dynamic force change, was employed recently (Georgopoulos et al., 1992). The following experimental arrangement allowed dissociation between the dynamic and static components of the force exerted. Monkeys produced pure force pulses on an isometric handle in the presence of a constant force bias so that the net force (i.e., the vector sum of the monkey's force and the bias force) was in a visually specified direction. The net force developed over time had to stay in the specified direction and to increase in magnitude to exceed a required intensity threshold. Now consider the case in which the directions of the net and bias forces differ, for example, by being orthogonal. For the task to be performed successfully under these conditions, the animal's force has to change continuously in direction and magnitude so that, at any moment during force development, the vector sum of this force and the bias force is in the visually specified direction. Thus, this experimental arrangement effectively dissociated the animal's force vector, the direction of which changed continuously in a trial from the net force vector, the direction of which remained invariant. Eight net force directions and eight bias force directions were employed. Recordings of neuronal activity in the motor cortex revealed that the activity of single cells was directionally tuned in the absence of bias force, and that this tuning remained invariant when the same net forces were produced in the presence of different directions of bias force. These results demonstrated that cell activity does not relate to the direction of the animal's force. Since the net force is equivalent to the dynamic component of the force exerted by the animal, after a static component vector (equal and opposite to the force bias) is subtracted, these findings suggest that the motor cortex provides the dynamic force signal during force development; other, possibly subcortical, structures could provide the static compensatory signal. This latter signal could be furnished by antigravity neural systems, given that most static loads encountered are gravitational in nature. According to this general view, the force exerted by the subject consists of dynamic and static components, each of which is controlled by different neural systems; these signals would converge in the spinal cord and provide an ongoing integrated signal to the motoneuronal pools.

Conclusion: Neural Mechanisms of Copying

Copying is a rich activity that involves coordinated interaction among several areas of the brain. In a continuation of our behavioral studies described previously, we have investigated the neural mechanisms underlying the copying function. For that purpose, we trained monkeys to copy simple geometrical shapes, including triangle, square, inverted triangle, and trapezoid. We then used a multielectrode system to record the activity of single cells in the periprincipalis region of the prefrontal cortex. We found the following (Averbeck et al., 2001; Chafee et al., 2001): Movement trajectories were segmented as evidenced by multiple positive-going zero-acceleration crossings. An analysis of covariance was used to assess the effect of various motor and shape variables on cell activity during the drawing of a segment. These variables included the serial position of the segment and shape drawn (as fixed factors), maximum segment speed, time to maximum speed, direction of segment in space, X-Y position of the segment's midpoint, average X-Y position of the eye during the drawing of a segment, and the sequential trial number. We found a significant relation to the serial position of the segment (46% of cells) and to the shape being drawn (51%). Smaller percentages of cells were related to other factors. Plots of adjusted neural activity means against the serial position of the segment revealed several types of systematic variation, including monotonically increasing and decreasing functions as well as parabolic functions.

These findings draw attention to the fact that usually simple shapes are drawn as a sequence of movements; in that respect, then, copying can be regarded as an instance of serially ordered behavior. A key idea in Lashley's formulation of the problem of serial order in behavior is the postulated neural representation of all serial elements before the action begins. We investigated this question by recording simultaneously the activity of small neuronal ensembles in prefrontal cortex during copying. As mentioned previously, the shapes were drawn as sequences of movement segments, and the drawing of these segments was associated with distinct patterns of neuronal ensemble activity. We found that these patterns were present during the time preceding the actual drawing. In fact, the rank of the strength of representation of a segment in the neuronal population during that time predicted the serial position of the segment in the motor sequence such that, for example, the segment with the highest strength was the first in the sequence, and so on. Lashley (1951) theorized that errors in motor sequences would be most likely to occur in the execution of elements that had prior representations of nearly equal strength. Now, in the previous analysis, the strengths of neural representation of middle segments in the drawing trajectory of a square, for example, were closer to each other than to the initial and last segments. A prediction based on this finding is that more errors would be committed in the drawing of middle segments in comparison to either the first or last segments. Indeed, more errors were committed during the drawing of the middle segments, and comparatively few were made during the drawing of either the initial or last segments. Specifically, over the entire data collection period, the monkey committed an average of 122, 249, and 165 errors during drawing of the initial, middle, and final segments, respectively. These findings connect the strength of the representation of the upcoming serial elements in copying to the actual performance and its variability with respect to errors committed. Overall, they provide additional evidence in support of Lashley's idea on the cotemporal representation of serial order and its effects on the serial behavior emitted subsequently. Finally, it is worth noting that this work has brought together the ideas of two great masters of the 20th century, Bernstein and Lashley, in the context of copying. And this is only one instance, among many others, of their sharp and inspiring ideas and intuitions in the field of motor behavior.

Acknowledgments

This work was supported by United States Public Health Service grants NS17413 and MH48185, the Department of Veterans Affairs, and the American Legion Chair in Brain Sciences.

References

Amirikian, B., and Georgopoulos, A.P. (2000) Directional tuning profiles of motor cortical cells. *Neurosci Res* 36: 73-79.

Ashe, J., Taira, M., Smyrnis, N., Pellizzer, G., Georgakopoulos, T., Lurito, J.T., and Georgopoulos, A.P. (1993) Motor cortical activity preceding a memorized movement trajectory with an orthogonal bend. *Exp Brain Res* 95: 118-130.

Averbeck, B.B., Chafee, M.V., Crowe, D.A., and Georgopoulos, A.P. (1998) Multidimensional scaling (MDS) of visual shapes and their motor copies. *Soc Neurosci Abstr* 24: 1180.

Averbeck, B.B., Chafee, M.V., Crowe, D.A., and Georgopoulos, A.P. (2001) Single unit activity related to serial order during the copying of geometric shapes. *Soc Neurosci Abstr* 467.2.

Behrmann, M., Winocur, G., and Moscovitch, M. (1992) Dissociation between mental imagery and object recognition in a brain-damaged patient. *Nature* 359: 636-637.

Benton, A.L. (1967) Constructional apraxia and the minor hemisphere. *Confin Neurol* 29: 1-16.

Bernstein, N. (1935) The problem of interrelation of co-ordination and localization. *Arch Biol Sci* 38. (Reproduced in Whiting, H.T.A. (Ed.) (1984) *Human motor actions*, pp. 77-119. North-Holland: Amsterdam.)

Caminiti, R., Johnson, P.B., and Urbano, A. (1990) Making arm movements within different parts of space: Dynamic aspects in the primate motor cortex. *J Neurosci* 10: 2039-2058.

Chafee, M.V., Averbeck, B.B., Crowe, D.A., and Georgopoulos, A.P. (2001) Motor sequence representation of prefrontal neurons predicts error patterns in drawing. *Soc Neurosci Abstr* 533.3.

Cipolotti, L., and Denes, G. (1989) When a patient can write but not copy: Report of a single case. *Cortex* 25: 331-337.

De Renzi, E. (1982) Disorders of space exploration and cognition. Wiley: Baffins Lane.

Fu, Q.-G., Flament, D., Coltz, J.D., and Ebner, T.J. (1995) Temporal coding of movement kinematics in the discharge of primary motor and premotor neurons. *J Neurophysiol* 73: 836-854.

Fu, Q.-G., Suarez, J.I., and Ebner, T.J. (1993) Neuronal specification of direction and distance during reaching movements in the superior precentral premotor area and primary motor cortex of monkeys. *J Neurophysiol* 70: 2097-2116.

Gainotti, G. (1985) Constructional apraxia. In Frederiks, J.A.M. (Ed.), *Handbook of clinical neurology*, pp. 491-506. Amsterdam: Elsevier.

Georgopoulos, A.P. (1995) Current issues in directional motor control. *Trends in Neurosci* 18: 506-510.

Georgopoulos, A.P., Ashe, J., Smyrnis, N., and Taira, M. (1992) Motor cortex and the coding of force. *Science* 256: 1692-1695.

Georgopoulos, A.P., Caminiti, R., Kalaska, J.F., and Massey, J.T. (1983) Spatial coding of movement: A hypothesis concerning the coding of movement direction by motor cortical populations. *Exp Brain Res* (Suppl 7): 327-336.

Georgopoulos, A.P., Kettner, R.E., and Schwartz, A.B. (1988) Primate motor cortex and free arm movements to visual targets in three-dimensional space: II. Coding of the direction of movement by a neuronal population. *J Neurosci* 8: 2928-2937.

Georgopoulos, A.P., Schwartz, A.B., and Kettner, R.E. (1986) Neuronal population coding of movement direction. *Science* 233: 1416-1419.

Hinton, G. (1984) Some computational solutions to Bernstein's problems. In Whiting, H.T.A. (Ed.), *Human motor actions,* pp. 413-438. Amsterdam: North-Holland.

Hocherman, S., and Wise, S.P. (1991) Effects of hand movement path on motor cortical activity in awake, behaving rhesus monkeys. *Exp Brain Res* 83: 285-302.

Kettner, R.E., Schwartz, A.B., and Georgopoulos, A.P. (1988) Primate motor cortex and free arm movements to visual targets in three-dimensional space. III. Positional gradients and population coding of movement direction from various movement origins. *J Neurosci* 8: 2938-2947.

Lashley, K.S. (1951) The problem of serial order in behavior. In Jeffress, L.A. (Ed.), *Cerebral mechanisms in behavior,* pp. 112-136. New York: Wiley.

Lukashin, A.V., Amirikian, B.R., Mozhaev, V.L., Wilcox, G.L., and Georgopoulos, A.P. (1995) Modeling of motor cortical operations by an attractor network of stochastic neurons. *Biol Cybern* 74: 255-261.

Lukashin, A.V., and Georgopoulos, A.P. (1993) A dynamical neural network model for motor cortical activity during movement: Population coding of movement trajectories. *Biol Cybern* 69: 517-524.

Lukashin, A.V., and Georgopoulos, A.P. (1994a) A neural network for coding of trajectories by time series of neuronal population vectors. *Neural Comput* 6: 19-28.

Lukashin, A.V., and Georgopoulos, A.P. (1994b) Directional operations in the motor cortex modeled by a network of spiking neurons. *Biol Cybern* 71: 79-85.

Lukashin, A.V., Wilcox, G.L., and Georgopoulos, A.P. (1994) Overlapping neural networks for multiple motor engrams. *Proc Natl Acad Sci* USA 91: 8651-8654.

Mack, J.L., and Levine, R.N. (1981) The basis of visual constructional disability in patients with unilateral cerebral lesions. *Cortex* 17: 515-532.

Piercy, M., Hécaen, H., and Ajuriaguerra, J. (1960) Constructional apraxia associated with unilateral cerebral lesions. Left and right sided cases compared. *Brain* 83: 225-242.

Salinas, E., and Abbott, L.F. (1994) Vector reconstruction from firing rates. *J Comp Neurosci* 1, 89-107.

Schwartz, A.B. (1994) Direct cortical representation of drawing. *Science* 265: 540-542.

Scott, S.H., and Kalaska, J.F. (1997) Reaching movements with similar hand paths but different arm orientations: I. Activity of individual cells in motor cortex. *J Neurophysiol* 77: 826-852.

Servos, P., and Goodale, M.A. (1995) Preserved visual imagery in visual form agnosia. *Neuropsychologia* 33: 1383-1394.

Shepard, R.N. (1980) Multidimensional scaling, tree-fitting, and clustering. *Science* 210: 390-398.

Whiting, H.T.A. (1984) *Human motor actions.* North-Holland: Amsterdam.

2

The Startle Reflex, Voluntary Movement, and the Reticulospinal Tract

J.C. Rothwell
Sobell Department, Institute of Neurology, London

J. Valls-Solé
Unitat d'EMG, Servei de Neurologia,
Hospital Clinic, Barcelona

The cerebral cortex, reticular formation, and vestibular nuclei all send descending motor fibers to the spinal cord. However, with the exception of the corticospinal tract (particularly its large-diameter component), there is remarkably little information on the function on these individual systems, especially in man. Here, we present data on possible methods to study the reticulospinal system in humans and, in particular, focus on possible roles of this system in normal voluntary movement.

The Startle Reflex

A large body of evidence from animal studies suggests that the startle response originates in the caudal brainstem. Lesioning experiments in the rat have implicated the median bulbopontine reticular formation (Szabo and Hazafi, 1965), particularly the nucleus reticularis pontis caudalis (Davis et al., 1982; Hammond, 1973; Leitner et al., 1980), as a primary center subserving the acoustic startle response. The efferent limb may be the reticulospinal and bulbospinal tracts that originate in this area (Schiebel and Schiebel, 1958; Torvik and Brodal, 1957). If the same organization operates in man, then the startle may be a method of analyzing reticulospinal function in humans.

The startle reflex in man consists of a generalized flexion response that habituates rapidly over two to six trials (Landis and Hunt, 1939). It may be generated by any form of stimulation but is usually provoked by a short, loud auditory tone. The absolute onset latency of the electromyographic (EMG) responses in the startle is rather variable and depends on the modality of the stimulus, the intensity of the stimulus, and the expectancy of the subject (Brown et al., 1991b; Chokroverty et al., 1992; Matsumoto et al., 1992; Wilkins et al., 1986). Typical onset latencies would be of the order of 60 ms in the sternocleidomastoid muscle and 75 ms in the biceps. Although the absolute latency of the responses is variable, the pattern of activity between different muscles is usually more stereotyped. Figure 2.1 shows the typical pattern of activity following an auditory startle in a single normal subject, recording from muscles around the neck, face, arms, and trunk.

The response appears to begin in the orbicularis oculi muscle. However, both Colebatch et al. (1990) and Brown et al. (1991b) have argued that this initial activity is not part of the true startle pattern but simply represents a blink reflex to the auditory stimulus. Three factors support this interpretation. First, it is sometimes possible to see a clear separation in the orbicularis oculi response between an early, short period of activity and a later, long-lasting period of activity. Second, the early component does not habituate as rapidly to repeated presentation of the auditory tone as the startle pattern. Thus, presentation of the auditory stimulus on the first occasion may give a full-blown startle pattern on the first trial, but after two or three presentations, the only remaining response is the initial portion of the orbicularis oculi burst. Third, EMG responses in the mentalis, another muscle innervated by the seventh cranial nerve, have a longer latency than those of the first component of the orbicularis oculi response that is not explicable by a longer peripheral conduction time. This latency fits well with the presumed caudo-rostral pattern of innervation of cranial-nerve-innervated muscles within the startle response detailed in the following section.

If the initial part of the orbicularis oculi response can be assumed to be a blink reflex, then the startle pattern proper begins with an activity in the sternocleidomastoid muscle. This is followed by activity in the mentalis and then the masseter muscles, and then by activity in the limb trunk muscles. The pattern and onset latencies of the responses in the startle are very different from those seen in the same muscles after transcranial electrical or magnetic stimulation of the motor cortex (Rothwell et al., 1991). Table 2.1 gives a comparison of the interval between the onset of EMG activity in the sternocleidomastoid and masseter or rectus abdominus (RA) muscles and the interval between activity in the biceps and abductor pollicis brevis (APB) muscles in normal subjects after a startle stimulus or after stimulation of the motor cortex. Three features are evident:

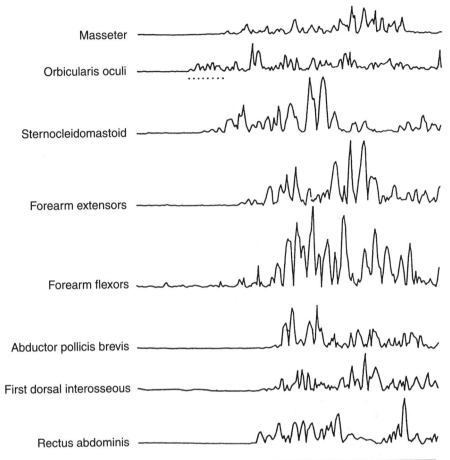

Masseter

Orbicularis oculi

Sternocleidomastoid

Forearm extensors

Forearm flexors

Abductor pollicis brevis

First dorsal interosseous

Rectus abdominis

Figure 2.1 Rectified EMG record of a single startle response recorded in a healthy individual. A 124-dB tone of 50 ms duration was delivered to both ears at the start of the trace. Excluding the auditory blink reflex (dotted line under the initial burst of the orbicularis oculi response), the first EMG activity is recorded in the sternocleidomastoid and then later in the masseter. There is a disproportionately long latency to the abductor pollicis brevis and the first dorsal interosseous. Horizontal calibration line, 50 ms; vertical calibration line, 0.05 mV. From Brown et al., 1991b.

1. In the startle response, there appears to be a caudo-rostral pattern of innervation of cranial-nerve-innervated muscles that begins in the sternocleidomastoid and spreads up to the mentalis (seventh nerve) and then masseter (fifth nerve) muscles.

2. The excess latency of the rectus abdominus response over the sternocleidomastoid is some three times longer in the startle than it is after stimulation of the motor cortex. Assuming that the peripheral conduction delay is the same

in both cases, then the conduction velocity of the efferent pathway within the spinal cord must be much slower for the startle than it is for the pathway activated by cortical stimulation. Since the latter is now thought to be the large-diameter component for the cortical spinal tract (Boyd et al., 1986), the slower conduction of the startle would be consistent with a reticulospinal projection.

3. In the startle, the latency to the onset of EMG responses in intrinsic hand muscles is disproportionately long, even when allowing for the slow conduction in the spinal efferent pathway. Again, this pattern is unlike the preferential accessibility of the hands seen after activation of the corticospinal system and indicates that a quite different pathway is used in the startle.

In summary, this electrophysiological evidence suggests that the physiological auditory startle reflex in man is mediated by an efferent system originating in the caudal brainstem. The spinal projections of this system are relatively slow-conducting and are distributed predominantly to axial muscles. The pattern of muscle recruitment in the response is determined by the distance of each segmental level from the caudal brainstem, with two exceptions. First, an auditory reflex precedes the

Table 2.1   A Comparison of the Efferent Pathways Subserving the Normal Auditory Startle Reflex With Corticobulbar and Pyramidal Pathways in Man

	Auditory startle reflex (ms)	Magnetic stimulation of motor cortex (ms)
Excess latency of masseter over SCM	0.7 ms	−2.9 ms
Excess latency of RA over SCM	24.0 ms	8.0 ms
Excess latency of APB over biceps	22.4 ms	10.5 ms

The differences in the latencies to onset of EMG activity between individual muscles in the normal auditory startle reflex are calculated from the median latencies from 12 healthy seated subjects. The stimulus was a 1-kHz, 124-dB tone of 50 ms duration delivered to both ears at intervals of about 20 min. The differences in the latency to onset of EMG activity between masseter and sternocleidomastoid (SCM), between sternocleidomastoid and rectus abdominis (RA), and between biceps and abductor pollicis brevis (APB), following magnetic stimulation of the motor cortex, are calculated from the mean latencies reported by Cruccu et al. (1989) and Thompson et al. (1991) in normal subjects. The pattern of recruitment of cranial nerves seen following magnetic stimulation of the motor cortex is reversed in the normal auditory startle reflex. The differences in latency to onset of EMG activity between sternocleidomastoid and rectus abdominis, and between biceps and abductor brevis, are longer in the auditory startle reflex than following magnetic stimulation of the motor cortex.

auditory startle reflex in the orbicularis oculi. Second, the latencies of intrinsic hand muscle responses are disproportionately delayed.

It should be noted that this general pattern of the startle reflex as described earlier may be complicated in some instances by one of the following mechanisms:

- Matsumoto et al. (1992) showed very clearly that the intensity of the startling stimulus could affect the latency of the response. This effect may differ from muscle to muscle, and therefore the intermuscle latencies could differ subtly and have different stimulation intensities.

- Particularly with startles elicited by somatosensory inputs, additional reflexes could be evoked that could complicate the startle pattern. For example, flexor reflexes might be produced by intense somatosensory stimuli, and the auditory blink reflex is elicited by acoustic inputs.

- Brown et al. (1991a) showed that the latency of startle reflex responses could vary according to the posture of the subject. For example, reflex activity in the tibialis anterior and soleus muscles is often absent or very rare in normal subjects when they are seated. However, reflexes can be recorded twice as frequently when standing and at a latency that is 40 to 60 ms shorter.

Pathology of the Startle Reflex

The idea that a startle reflex might arise in the caudal brainstem and utilize slowly conducting reticulospinal pathways is consistent with several pathological studies in man. Brown et al. (1991c) studied several patients with hyperekplexia, a condition characterized by excessive sensitivity and lack of habituation to startle stimuli. Some patients with hereditary forms of the disease are known to have a mutation in the glycine receptor gene (Shiang et al., 1993) and also show abnormalities in some spinal reflexes, as well as in the startle response (Floeter et al., 1996). Although the startle was more readily elicited in these patients than in normal subjects, the pattern of activation was the same in both cases. Several symptomatic cases had pathology affecting the brainstem (Brown et al., 1991c).

In a further study, Vidailhet et al. (1992) showed that the startle reflex was virtually absent in patients with progressive supranuclear palsy. Although the pathology in this condition is widespread, one area of particularly high neuronal loss is the lower pontine reticular formation.

Interaction Between the Startle Reflex and Voluntary Motor Responses

Voluntary motor responses to sensory stimulation usually have a much longer latency than those of the startle response. For example, in arm muscles, a startle reaction may have a latency less than 80 ms. In contrast, voluntary reaction time to a visual "go" signal is on the order of 150 ms. The reaction time to auditory and somatosensory stimuli is shorter, but even then is rarely less than 100 ms (Thompson et al., 1992). The difference in latency between the voluntary and startle responses is thought to be due to the fact that when subjects move voluntarily in response to a reaction signal, the cerebral cortex must play a role in identifying the sensory stimulus and releasing the instructions to move. In contrast, startle reactions are thought to be reflex phenomena caused by sensory inputs to the reticular formation of the brainstem activating descending reticulospinal tract projections to the spinal cord.

Recently, Valls et al. (1995) showed that voluntary reaction times can be considerably reduced if a very loud startling sound is given at the same time as the visual "go" signal. The degree of shortening is much greater than that observed in conventional intersensory facilitation (Nickerson, 1973) and presumably represents a specific startle-related effect. The question is, What neural mechanisms are responsible for this phenomenon? Since the reaction times are the same as those of the startle reaction itself, the simplest explanation is that the activity observed in these experiments consists of two components: an early startle reflex and a late voluntary response, the true onset of which is masked by the startle. There is an alternative explanation, however. It could be that the high-intensity acoustic stimulus somehow releases the movement being prepared for voluntary execution at a speed far quicker than usual, perhaps bypassing cortical circuitry and activating muscles using the same pathways as the startle reaction itself.

We have therefore conducted experiments to test these two competing explanations for startle-induced reaction time shortening (Valls et al., 1999). We examined the effects of the startling stimulus on two stereotyped EMG patterns of activity that are generally thought to be generated in the central nervous system. One was the triphasic agonist/antagonist/agonist burst of a rapid, self-terminated wrist flexion or extension movement (Hallett et al., 1977). The other was the structured EMG pattern that accompanies the action of rising on tiptoes from the standing position (Nardone and Schieppati, 1988). This begins with a silence of the soleus muscle and activation in the tibialis anterior, both of which tend to bring the center of gravity forward. This is followed by a large burst of activity in the soleus muscle that pushes the heel from the ground. Both the triphasic ballistic movement pattern and the postural response are thought to be centrally programmed and can be generated without that significant contribution from peripheral afferent inputs (Forget and Lamarre, 1990; Sanes and Jennings, 1984). We postulated that if a loud acoustic stimulus shortens reaction time by superimposing a startle reflex onto the onset of the voluntary response, there would be a disruption of the stereotyped EMG pattern for both movements. Conversely, if the startling stimulus somehow bypasses some of the normal reaction time circuitry, the response would be sped up with no alteration of the EMG pattern.

Figures 2.2 and 2.3 show the EMG patterns produced when subjects react voluntarily to the onset of a visual cue and when the same subjects receive an unexpected, loud startling stimulus at the same time as the visual cue. In both cases, the

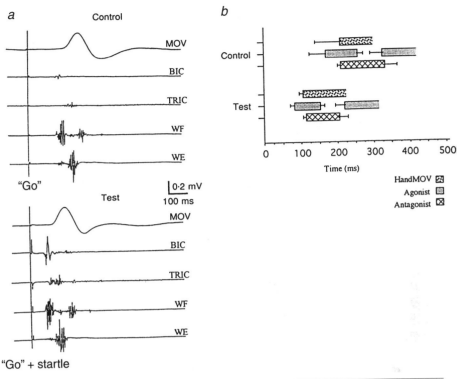

Figure 2.2   Raw data example and mean data from all subjects in the arm movement trials. Recordings from a single subject performing a ballistic wrist flexion movement. *(a)* Control trials in which the subject responded only to the visual cue and Test trials in which the visual cue was accompanied by a loud startling auditory stimulus. MOV = displacement of the wrist joint; BIC = biceps brachii; TRIC = triceps brachii; WF = flexor carpi radialis; WE = extensor carpi radialis. *(b)* Schematic representation of the mean EMG pattern of all subjects in the Control and Test trials. The left extent of the bars represents the mean onset latency, whereas the horizontal line shows one standard deviation of the mean. The length of the bars represents the duration of the EMG bursts, and the horizontal lines at the right side of the bar show one standard deviation of this mean. Duration was only measured for the first agonist and the antagonist bursts. Note that the pattern of the EMG bursts (i.e., the interburst interval and the burst durations) is the same in the Control and Test trials, even though the onset latency is substantially reduced in the Test trials.

From Valls et al., 1999.

effect of the startling stimulus is to speed up the voluntary reaction without affecting the form of its response in any way. Thus, in the ballistic wrist movements, the relative timing and amplitude of activity of the triphasic pattern in the flexor and extensor muscles remain unchanged, but the onset of activity in both muscles is considerably reduced. The same is true of the postural task. Reaction to a visual cue is accompanied by an initial silence in the soleus, followed by a burst of activity in

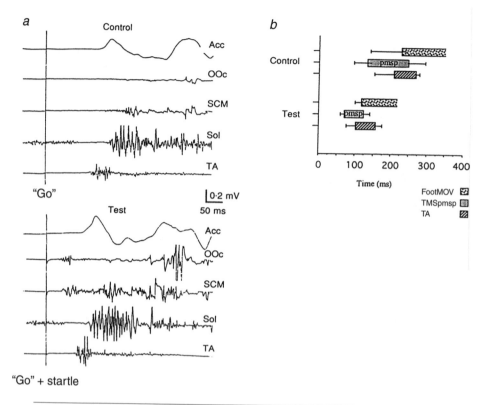

Figure 2.3   Raw data example and mean data from all subjects when performing the rise onto tiptoes task. *(a)* Recordings from a single subject performing a sudden rise onto tiptoes in Control and Test trials. Acc = accelerometric recording from the dorsum of the foot; OOc = orbicularis oculi; SCM = sternocleidomastoid; Sol = soleus; TA = tibialis anterior. *(b)* Schematic representation of the mean pattern of EMG activity from all subjects observed in the leg movement task. The left extent of the bars represents the mean onset latency of the events, whereas the horizontal line shows one standard deviation of the mean. The length of the bars represents the duration of the events, and the horizontal lines at the right side of the bar show one standard deviation of this mean. pmsp = premovement silent period. Duration was only measured for the TMSpmsp and the TA burst. Note the preservation of the burst durations and interburst latencies despite the shortening of movement onset.
From Valls et al., 1999.

the tibialis anterior and then a final burst of activity in the soleus. The relative timing of the onset of silence, the onset of activity in the tibialis anterior, and the final soleus activity are all unchanged by the addition of the startling auditory stimulus. The main effect is simply to shorten the latency of the whole movement pattern.

   The conclusion from these experiments is that a loud startling stimulus speeds up the execution of a voluntary movement without changing its main characteristics. However, it should be noted that some aspects of the startle reflex do persist under these conditions. For example, in figure 2.3, EMG records from the orbicularis oculi and the sternocleidomastoid muscle are also included in the raw data from one subject. These traces show that when an auditory startling stimulus is given, bursts of activity occur at very short latency in the orbicularis oculi and sternocleidomastoid that we assume are part of the normal startle response. However, no such bursts of activity are seen in the leg muscles, where the voluntary pattern seems to override any true startle that might have occurred.

We (Valls et al., 1999) have argued that the reaction time shortening was unlikely to be due to the phenomenon of intersensory facilitation between the visual "go" signal and the startling acoustic stimulus (Nickerson, 1973). Intersensory facilitation may shorten reaction times by 20 to 50 ms, but the amount of shortening seen in the present experiments was generally more than 70 ms. Thus, although intersensory facilitation may have played some role in speeding the reactions, other factors must also have been involved.

We can only speculate on the mechanism that could speed up voluntary responses by so great an amount. However, the calculations suggest that with onset latencies of 65 ms for forearm muscle responses, or even 60 ms for the onset of the soleus silent period during standing on tiptoes, there is relatively little time for transmission of responses through cortical circuitry. Indeed, unless cortical activation occurred via an unknown and very rapid input from auditory afferents to the motor cortex, it is unlikely that the fastest reactions observed during the auditory startle stimulus could have been produced by activity in conventional corticospinal systems.

Since elements of the startle reflex proper were elicited in muscles not involved in the voluntary reaction, the simplest explanation is that the voluntary reaction was driven at the speed of the startle reaction while maintaining features of the voluntary motor program. Another possibility is that the interaction between startle input and voluntary output occurred in the reticular formation where the startle response originates. The implication is that under certain conditions, the motor program that is being prepared for voluntary execution is triggered by activation of the same reticular structures that are responsible for the startle reflex. Preparatory premovement activity occurs in parallel at many levels of the nervous system, from cortex to spinal cord. For example, the spinal monosynaptic reflex changes before movement as well as in the preparatory period before the reaction signal is given. Motor-evoked potentials to transcranial magnetic stimulation also change both in the reaction and preparatory periods before voluntary movement (Rossini et al., 1988). Indeed, even the startle pattern itself can be modulated by the stance of the subject before the stimulus is given (Brown et al., 1991a). Our hypothesis is that sufficient detail of the voluntary movement characteristics may be stored in brainstem and spinal centers so that, on occasion, the whole motor program can be triggered without the expected command from the cerebral cortex. This is not to deny the cerebral cortex a crucial role in preparing the response parameters or to exclude it from contributing to later parts of the response. However, the implication is that the cerebral cortex does not need to play the lead role in initiating voluntarily prepared responses.

Conclusion

Much recent work has concentrated on the corticospinal system and its involvement in control of distal manipulative movements of the hand and fingers. In this chapter we alert researchers to the importance of other descending fibers in movement control. The classic experiments of Lawrence and Kuypers (1968) showed that total surgical section of the pyramidal tracts in the monkey had little effect on most limb and trunk movement. Animals could sit, run around the cage, and climb up its bars with little problem; their main deficit was in independent finger movement. The conclusion is that tracts other than the corticospinal tract contribute greatly to volitional control of many nonmanual tasks.

Here we suggest that the reticulospinal tracts may form an important parallel connection between the cerebral cortex and spinal cord. These tracts receive inputs from the cortex and can be studied using the startle reflex in intact humans. Volitional inputs to these systems are likely to be responsible for the rapid onset of startle-induced voluntary reactions. It will be interesting in future studies to test the role of such systems in recovery from function after damage to direct corticospinal systems.

References

Boyd, S.G., Rothwell, J.C., Cowan, J.M., Webb, P.J., Morley, T., Asselman, P., et al. (1986) A method of monitoring function in corticospinal pathways during scoliosis surgery with a note on motor conduction velocities. *J Neurol Neurosurg Psychiatry* 49: 251-257.

Brown, P., Day, B.L., Rothwell, J.C., Thompson, P.D., and Marsden, C.D. (1991a) The effect of posture on the normal and pathological auditory startle reflex. *J Neurol Neurosurg Psychiatry* 54: 892-897.

Brown, P., Rothwell, J.C., Thompson, P.D., Britton, T.C., Day, B.L., and Marsden, C.D. (1991b) New observations on the normal auditory startle reflex in man. *Brain* 114: 1891-1902.

Brown, P., Rothwell, J.C., Thompson, P.D., Britton, T.C., Day, B.L., and Marsden, C.D. (1991c) The hyperekplexias and their relationship to the normal startle reflex. *Brain* 114: 1903-1928.

Chokroverty, S., Walczak, T., and Hening, W. (1992) Human startle reflex: Technique and criteria for abnormal response. *Electroencephalogr Clin Neurophysiol* 85: 236-242.

Colebatch, J.G., Barrett, G., and Lees, A.J. (1990) Exaggerated startle reflexes in an elderly woman. *Mov Disord* 5: 167-169.

Cruccu, G., Berardelli, A., Inghilleri, M., and Manfredi, M. (1989) Functional organization of the trigeminal motor system in man. A neurophysiological study. *Brain* 112: 1333-1350.

Davis, M., Gendelman, D.S., Tischler, M.D., and Gendelman, P.M. (1982) A primary acoustic startle circuit: Lesion and stimulation studies. *J Neurosci* 2: 791-805.

Floeter, M.K., Andermann, F., Andermann, E., Nigro, M., and Hallett, M. (1996) Physiological studies of spinal inhibitory pathways in patients with hereditary hyperekplexia. *Neurology* 46: 766-772.

Forget, R., and Lamarre, Y. (1990) Anticipatory postural adjustment in the absence of normal peripheral feedback. *Brain Res* 508: 176-179.

Hallett, M., Shahani, B.T., and Young, R.R. (1977) Analysis of stereotyped voluntary movements at the elbow in patients with Parkinson's disease. *J Neurol Neurosurg Psychiatry* 40: 1129-1135.

Hammond, G.R. (1973) Lesions of pontine and medullary reticular formation and prestimulus inhibition of the acoustic startle reaction in rats. *Physiol Behav* 10: 239-243.

Landis, C., and Hunt, W.A. (1939) *The startle pattern.* New York: Farrar and Reinhart.

Lawrence, D.G., and Kuypers, H.G.J.M. (1968) The functional organisation of the motor system in the monkey. I. The effects of bilateral pyramidal lesions. *Brain* 91: 1-14.

Leitner, D.S., Powers, A.S., and Hoffman, H.S. (1980) The neural substrate of the startle response. *Physiol Behav* 25: 291-297.

Matsumoto, J., Fuhr, P., Nigro, M., and Hallett, M. (1992) Physiological abnormalities in hereditary hyperekplexia. *Ann Neurol* 32: 41-50.

Nardone, A., and Schieppati, M. (1988) Postural adjustments associated with voluntary contraction of leg muscles in standing man. *Exp Brain Res* 69: 469-480.

Nickerson, R.S. (1973) Intersensory facilitation of reaction time: Energy summation or preparation enhancement? *Psychol Rev* 80: 489-509.

Rossini, P.M., Zarola, F., Stalberg, E., and Caramia, M.D. (1988) Premovement facilitation of motor evoked potentials in man during transcranial stimulation of central motor pathways. *Brain Res* 458: 20-30.

Rothwell, J.C., Thompson, P.D., Day, B.L., Boyd, S., and Marsden, C.D. (1991) Stimulation of the human motor cortex through the scalp. *Exp Physiol* 76(2): 159-200.

Sanes, J.N., and Jennings, V.A. (1984) Centrally programmed patterns of muscle activity in voluntary motor behavior of humans. *Exp Brain Res* 54: 23-32.

Schiebel, M.E., and Schiebel, A.B. (1958) Structural substrates for integrative patterns in the brainstem reticular core. In Jasper, H.H, Proctor, L.D., Knighton, R.S., Noshay, W.C., and Costello, R.T. (Eds.), *Reticular formation of the brain,* pp. 31-55. Boston: Little, Brown.

Shiang, R., Ryan, S.G., Zhu, Y.Z., Hahn, A.F., O'Connell, P., and Wasmuth, J.J. (1993) Mutations in the alpha 1 subunit of the inhibitory glycine receptor cause the dominant neurologic disorder, hyperekplexia. *Nat Genet* 5: 351-358.

Szabo, I., and Hazafi, K. (1965) Elicitability of acoustic startle after brainstem lesions. *Acta Physiol Acad Sci Hung* 27: 155-165.

Thompson, P.D., Colebatch, J.G., Brown, P., Rothwell, J.C., Day, B.L., Obeso, J.A., et al. (1992) Voluntary stimulus-sensitive jerks and jumps mimicking myoclonus or pathological startle syndromes. *Mov Disord* 7: 257-262.

Thompson, P.D., Day, B.L., Crockard, H.A., Calder, I., Murray, N.M., Rothwell, J.C., and Marsden, C.D. (1991) Intra-operative recording of motor tract potentials at the cervico-medullary junction following scalp electrical and magnetic stimulation of the motor cortex. *J Neurol Neurosurg Psychiatry* 54: 618-623.

Torvik, A., and Brodal, A. (1957) The origin of reticulospinal fibres in the cat: An experimental study. *Anat Rec* 128: 113-137.

Valls, S.J., Rothwell, J.C., Goulart, F., Cossu, G., and Munoz, E. (1999) Patterned ballistic movements triggered by a startle in healthy humans. *J Physiol Lond* 516: 931-938.

Valls, S.J., Sole, A., Valldeoriola, F., Munoz, E., Gonzalez, L.E., and Tolosa, E.S. (1995) Reaction time and acoustic startle in normal human subjects. *Neurosci Lett* 195: 97-100.

Vidailhet, M., Rothwell, J.C., Thompson, P.D., Lees, A.J., and Marsden, C.D. (1992) The auditory startle response in the Steele-Richardson-Olszewski syndrome and Parkinson's disease. *Brain* 115: 1181-1192.

Wilkins, D.E., Hallett, M., and Wess, M.M. (1986) Audiogenic startle reflex of man and its relationship to startle syndromes. A review. *Brain* 109: 561-573.

Posturo-Kinetic Capacity and Postural Function in Voluntary Movements

Simon Bouisset and Serge Le Bozec
Laboratoire de Physiologie du Mouvement,
Université Paris-Sud

This chapter addresses the relations between anatomical body structure and postural function associated with voluntary movements. It is divided into three parts. In the first part, we develop some general considerations that are based on the application of Newton's laws to the human body. In particular, the role played by the support reaction forces is stressed and the concept of stability defined. In the second part, we present reasons why the voluntary movement can be considered as a perturbation of body balance. The anticipatory postural adjustments (APAs) are considered and their relations to postural programming explained. The third part is devoted to posturo-kinetic capacity (PKC), which is the ability to develop the necessary counterperturbation to the task-movement–related perturbation.

To assess the influence of postural conditions on PKC, experimental results dealing with voluntary segmental movements, performed on various support bases and by normal and disabled subjects, have been analyzed. To this end, APAs, task performance, and muscular synergy are considered. The analyses show that PKC depends not only on body balance stability but also on another postural parameter: the dynamic mobility of the postural chain.

The postural chain dynamic mobility results from anatomical (range of individual joint movements, geometrical configuration of body segments, etc.) and physiological (muscular contraction, stiffness, etc.) factors. It is proposed that body balance stability refers to preservation of global body balance, whereas postural chain mobility is related to local balance stability. Both PKC factors should be taken into account in motor act programming.

The conclusion concerns the "static" and "dynamic" roles played by the anatomical body structures in postural function associated with voluntary movement. The relations between posture and voluntary movement are complex and apparently conflicting: each voluntary movement perturbs balance, but imbalance must be controlled and to a certain extent limited so that the movement is performed efficiently. Therefore, control of postural phenomena has been considered a key condition for efficient performance and studied over the past two decades in a variety of motor tasks. But even if the amount of results is impressive, many questions remain unsolved.

This chapter focuses on the factors that govern postural chain behavior. To this end, the relations between anatomical body structure and the postural function in voluntary movement are considered an interactive dynamic process. More precisely, the relations between the focal and the postural chain are envisioned in a more synthetic context, within the framework of posturo-kinetic capacity. In other words, the problem is posed in terms of biomechanics that proceed from Newton's laws, insofar as they allow defining the causal link between forces and the movements they induce.

The Body As a Newtonian Structure

Newton's laws apply to all categories of mechanical structure—to inanimate physical as well as to living structures. Humans and animals must comply with the constraints resulting from Newton's laws. These constraints have been taken into account during the evolutionary process; they are taken into account at each instant of a motor act. To examine the influence of movement on posture, the body structure is considered as a system of articulated rigid solids interacting with its environment.

Global Body Movement and Support Reaction Forces

A direct consequence of Newton's laws is that global human body movements only result from forces that are external to the body. On the ground track, these forces are often limited to gravity and support reaction forces. One is constant; the other, which originates from the ground, and more generally from the physical support, varies as a function of the forces developed during the motor act. As Aristotle (1973) so clearly formulated long ago: "Animals that move need an external support in order to perform this movement, even for breathing."

The skeletal structure allows modeling of the human body as a system of rigid solids, termed *links* (Dempster, 1955). These links are movable in relation to each other and form an articular chain. From an anatomical viewpoint, the mobility of an articular chain is a function of the range of individual joint movements. However, data on isolated joints cannot entirely reflect the cumulative geometric background for the mobility of an articular chain; it is also important to know the relative orientation of a joint to the adjacent joints. The functional effect of mobility on postural function will be considered in the last section of the chapter.

In addition, the fact that the body segments are approximately rigid implies that forces (and torques) are transmitted without any time lag, and step by step, from the segment(s) the subject mobilizes intentionally to others and from the distal one(s) to the ground. This necessary transmission was clearly stated by E.J. Marey (1883) in presenting the first force-plate data: "When a muscular act results in a center of gravity elevation, reactions are transmitted from step to step onto our inferior extremity, and provoke a positive pressure increase on the dynamometer (the force-plate)."

Therefore, a clear understanding of the postural phenomena associated with voluntary movement necessitates taking into account the external forces, particularly the support reaction forces. This necessity leads to consideration of the support base.

Support Base and Support Reaction Forces

The support base, also named the underpropping area, is defined by the body areas in contact with the physical support; it is the interface between the body and the physical support. Biomechanical principles indicate that the reaction forces applied to the body, as well as the center of pressure (COP) displacement, depend on the support base characteristics.

On the one hand, the support reaction forces are in relation to the physical support characteristics. They depend on the properties of the materials (elasticity, friction, etc.) and on geometrical (flatness or curvature, inclination, etc.) and design (steadiness or oscillations, translations, etc.) support factors. For example, the support reaction forces are limited in amplitude by the elasticity and in direction by the friction of the support; when the external forces are limited to gravity and support reaction forces, the same is true for the center of gravity (COG) acceleration, insofar as ground forces and COG acceleration are equal within the subject's mass. Also, the support reaction forces are modified if a mechanical perturbation is applied to the physical support.

On the other hand, the support reaction forces are a function of the body areas in contact with the physical support. There are many possibilities: the subjects being able to stand on tiptoes, on heels, or on the entire foot, to stand on one foot or both feet, to sit with a more or less complete ischiofemoral contact, etc. The support base can remain constant, as in upper limb or trunk movements; change transiently, as in lower limb movements; or change periodically, as in locomotion. For a given task, the support reaction forces tend to increase if the support areas are reduced.

Also, the perimeter of the support base, which results from its configuration, defines the postural stability limits, as will be analyzed later. It also delimits, by construction, the COP displacement. As a consequence, for the motor tasks that require maintenance of a single support base, the COP fluctuates both in front of and behind the line of gravity (figure 3.1), providing accelerative and decelerative forces to control the COG movement (Murray et al., 1967). Therefore, the COG horizontal acceleration, being proportional to the difference between the COP and the COG horizontal displacements (Brenière et al., 1987), is limited by the boundaries of body contact area.

Finally, within a given support base configuration, the COG position varies according to the subject's posture, depending on whether the subject is upright or inclined forward, backward, sideways, etc., which also influences the stability conditions as well as COG horizontal acceleration. Therefore, the interface between the body and the physical support appears a priori to be an important constraint for body dynamics.

Muscular Forces and Support Reaction Forces

When the human body is considered as a whole, the muscular forces are internal to the mechanical system. As such, they cannot determine the overall movement of the

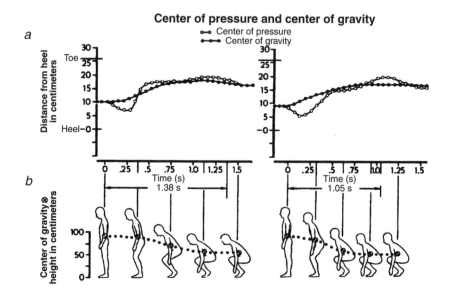

Figure 3.1   Center of gravity and center of pressure displacements during descent to squat posture. *(a)* Graphs show fore-aft excursions for free speed (left) and fast speed (right) descent. *(b)* Graphs show vertical trajectory of the center of gravity.

Reprinted from Murray, Seireg, and Scholz 1967.

body; they can only provoke segmental movements. In other words, if a muscular contraction provokes an articular movement, an upper limb flexion for instance, the rest of the body has a tendency to move in the opposite direction, in accordance with the action and reaction law (figure 3.2). On the ground track, the global center of gravity can only move if the body is in contact with a resistance originated from the environment, usually a physical support that offers the appropriate reaction, in addition to gravity: global body movements (i.e., the movement of the body center of gravity and in relation to it) result from the interaction between these two categories of forces.

Even though they are internal forces, however, the muscular forces play a major role (Bouisset and Maton, 1995). Provided that there is an external reaction to their action, the rotation of body segments they produce leads, in turn, to movements of the center of gravity (or around it). In other words, voluntary movement occurs as the result of a combination of muscle torques and ground reactions. As stated by Gray (1968), "if the body of an animal is at rest relative to its environment, it can only be set in motion by the application of an external force, and, consequently, if an animal is to move its body by its own unaided efforts, it must elicit a force from its external environment; similarly, an external environment force must be brought into play if an animal is to change its speed or direction of motion."

Also, as the muscular forces are under central nervous system (CNS) control, they are the only ones that can be used to manage the fatal effects of the external

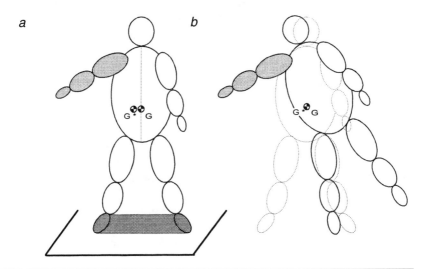

Figure 3.2 Shoulder abduction in two support base conditions. *(a)* The feet are in contact with the ground; there is a displacement of the body's COG. *(b)* The feet are no longer in contact with the ground, the upper limb and the rest of the body move in opposite directions, and the body's COG does not move (except if there is a residual gravity force). G is the initial position of the body's COG, which moves only when the feet are in contact with the ground; G* is the final position.

forces. Therefore, the effects of internal and external forces can be harmonized in such a way that the performance fulfills the task-movement constraints. That is why a complete understanding of the postural phenomena associated with voluntary movement necessitates consideration of the human body as being in interaction with the environment (i.e., taking into account the external as well as the muscular forces).

To examine the influence of movement on posture in this context, the system under study is divided into the following three parts, including

1. the articular chain, including segments that are moved voluntarily, which constitutes the "focal chain";

2. the rest of the body, in particular, the segments located between the segments that are voluntarily mobilized and the support surface(s), which is called the "postural chain"; and

3. the physical support(s) that is (are) in contact with the body part(s).

Body Posture, Equilibrium, and Stability

According to mechanics, an equilibrium state can be stable or unstable; it is termed *stable* when the structure that is initially in an equilibrium or a stationary state returns to its initial state after a perturbation. Hence, such states are dependent on the capacity to develop the forces necessary to oppose the perturbing forces: the greater the stability, the greater the forces able to perturb the initial state. Between its initial and final positions (i.e., between the onset and the end of a motor act), the structure is in a state of transient disequilibrium, which is termed *dynamic equilibrium*.

Human body balance stability must comply with two sets of requirements. The first are global requirements: they are the same as if the body were an invariant structure (i.e., as if the body segments' configuration were mechanically constrained). They result in the usual but incomplete contention according to which general body balance is secured when the projection on the ground of the center of gravity is situated within the support base perimeter. The second are local requirements: they result from the possible deformation of the body structure, which implies that each body part must also be balanced with respect to the underlying ones. Due to the configuration of the joint contact surfaces and the shape of bones, purely mechanical factors cannot fulfill this requirement. That is why appropriated muscular forces, originating from equilibration reactions, are necessary even in natural upright posture.

Hence, the stability limits of a system of articulated solids are exceeded when the projection of the center of gravity onto the ground falls outside the support base perimeter, or equivalently, if the ground reaction force vector is not located within the stability cone (McCollum and Leen, 1989). Of course, different configurations of the articulated chain can meet this requirement, but the stability limits are an invariant of the system under consideration. In humans, they depend on the support base characteristics and the subject's morphology and are under the control of postural reactions. For instance, in the natural upright posture, they are a function of the center of gravity height, the foot dimensions and width of stance, and the support surface friction.

Within the stability limits, a given posture is stable as long as the internal forces are sufficient to oppose the perturbing ones. In other words, there is a stability continuum: stability levels range from maximal to minimal, which corresponds to the instability threshold. For a given set of postural conditions, the stability level is a function of the perturbation intensity and can be evaluated by the phenomena involved in recovering the initial posture; for a given perturbation, the stability level depends on the postural conditions, particularly on the support base characteristics.

These considerations were helpful for analyzing postural maintenance, such as quiet sustained postures where, in addition to the gravitational forces, the perturbation comes from internal forces such as those originating in the firing of motor units and the respiratory and cardiac cycles. *Stability limit* (or "area of stability") as well as *stability level* (also called "steadiness") have been assessed in various standing postures in both normal and pathological adults or infants (e.g., Allum, 1990). They are defined by postural sway characteristics; one is the distance between the COP extreme positions, and the other is the COP excursion for a given time period (Murray et al., 1975). Insofar as COP is the expression of the motor signals that control the COG, larger COP excursions are often equated with a lesser stability level, which is a poorer balance control: for example, wide stance is considered more stable than narrow stance, which is more stable than one-legged stance. Therefore, it seems reasonable to assume that in postural maintenance, a greater COP excursion means a lower capacity to mobilize additional forces to oppose the internal perturbing ones. However, it does not mean that this capacity is necessarily reduced when an external perturbation is applied to the body.

The same general considerations can be used for analyzing postural stabilization under stance perturbations resulting from support translations, which have been extensively studied since Nashner (1976). In particular, it has been reported that there is a continuum of strategies, ranging from the "ankle strategy" to the "hip strategy" (Horak and Nashner, 1986), for compensating the imbalance resulting from the perturbation intensity or the support base length. Obviously, these strategies are only valid for a given initial posture and under the condition that the motor task requirements imply the preservation of the same support base.

Voluntary Movement As a Perturbation to Posture

The forces (and torques) that are developed during the movement are transmitted through the articular chain, including the segments located between the segment(s) that is (are) voluntarily mobilized and the physical support surface(s). Therefore, if "posture denotes first the shape of the body, that is, the geometrical relationship of the different parts to each other" (Martin, 1967), segmental as well as gross body movements tend to perturb posture and hence equilibrium, given that an equilibrium state is associated with a posture. But the relation between posture and equilibrium is not bi-univocal, as different body postures can result in the same COG position.

Postural Adjustments and Task-Movement Parameters

Neurologists have suggested, at least since Babinski's (1899) observations, that neurophysiological mechanisms underlie the postural reactions associated with voluntary movement. During the past two decades, biomechanical and electromyographic methods have made it possible to confirm and deepen neurological data and to extend the research to more fundamental questions.

Diverse experimental situations have been studied. Some are segmental voluntary movements, performed on a single support base, such as shoulder flexion or extension (Belenkii et al., 1967; Bouisset and Zattara, 1981, 1987, 1990; Clément et al., 1984; Horak et al., 1984; Lee, 1980; Lee et al., 1987; Maki, 1993; Riach and Hayes, 1990), shoulder horizontal (Aruin and Latash, 1995a) or lateral abduction (Vernazza et al., 1996), elbow flexion (Benvenuti et al., 1990; Friedli et al., 1984, 1988), wrist flexion (Hellebrandt et al., 1956), pushing or pulling a handle (Brown and Frank, 1987; Cordo and Nashner, 1982; Dick et al., 1986; Inglin and Woollacott, 1988; Lee et al., 1987; Wing et al., 1997), or trunk bending (Crenna et al., 1987; Houtz and Fisher, 1961; Oddsson and Thorstensson, 1986; Pedotti et al., 1989). Others are still segmental movements, but the support base is transitory, as in lower limb flexions/extensions (Do et al., 1991; Houtz, 1964; Nouillot et al., 1992; Rogers and Pai, 1990) or abductions (Mouchnino et al., 1991). Still others involve whole-body movement, which induces transitory support base transfer, such as elevation on tiptoes (Houtz and Fisher, 1961; Lipshits et al., 1981; Nardone and Schieppati, 1988), rocking on the heels (Nardone and Schieppati, 1988), kicking movements (Béraud and Gahéry, 1997), or whole-body reaching (Stapley et al., 1998) and lifting (Toussaint et al., 1998) tasks. In these tasks, even if the support base is transferred, there is still no body progression. Finally, in motor tasks such as locomotion, there is a periodical support base transfer that corresponds to a succession of balance losses and recoveries (Brenière et al., 1987).

The primary result of these studies is that postural adjustments associated with voluntary movement proceed from an active process, insofar as they result from muscular contractions, in addition to the passive articular forces that are transferred from one segment to the other according to the action and reaction law. More precisely, these studies point to several general features:

1. Motor command is distributed to postural as well as to focal segments.
2. Postural function can be carried out by any muscle and is not limited to the muscles that are termed postural on the basis of their fiber type content.
3. Postural activities are organized according to a well-defined sequence, which is task-specific.

4. The muscle excitation can be either tonic or phasic, and postural segments fixation can be a dynamic process.

5. Postural muscular activities usually precede, accompany, and follow the intentional movement.

6. Postural activities are adaptable insofar as they are modified in fatigue and conditioning, growth and old age, microgravity, etc.

7. Postural activities are modified, or even suppressed, in impairments of the nervous, muscular, and osteoarticular systems.

Therefore, postural phenomena associated with voluntary movement are claimed to be under CNS control.

Anticipatory Postural Adjustments and Postural Programming

According to Bernstein (1935), postural adjustments are assumed to constitute a part of the motor program. In other words, there is believed to be a "postural task" associated with the focal task. To identify the task parameters that are programmed, the anticipatory postural adjustments constitute a valuable tool.

Indeed, as the APAs precede by definition the onset of focal movement, they cannot result from reafferentation triggered by the focal movement; they are "preprogrammed." As the APAs occur in the postural chain, the postural chain can be said to be programmed. If APAs vary in relation to task parameters, it can be assumed that the postural chain is programmed in relation to these parameters. This reasoning is clearly supported by the assumption that anticipatory, contemporary, and consecutive postural adjustments are parts of the same motor program. In other words, it is assumed that as the postural process results from the movement process (the former would be the mathematical transform of the latter), and given that the movement process is continuous, a part of the postural adjustments, such as the APAs, can be supposed to be representative of the entire postural process. This assumption is compatible with theories on motor planning and is supported by experimental evidence; for example, the kinematics of postural segments, including anticipatory kinematics, display a continuous and reproducible pattern, and the amplitude of their extrema varies as a function of the task-movement parameters (Bouisset et al., 2000).

The present results are based mainly on the postural adjustments associated with upper limb voluntary movements performed by standing and sitting subjects on a single support base. This type of paradigm, which appears to be the simplest one, entails several advantages. The consideration of upper limb movements makes it easier to distinguish between the body part that is voluntarily moved and the rest of the body (i.e., between the focal and the postural chain). The continuous support base allows manipulation of movement parameters and interpretation of the results without any risk of misinterpretation related to support base effects. Moreover, as it is a very simple paradigm, it is supposed that the postural mechanisms described can guide the interpretation of more complex experimental situations.

It has been established that, like the movement itself, APAs associated with a voluntary movement are dynamic phenomena. They are specific to task parameters. Among these are the forthcoming movement velocity (Bouisset et al., 2000; Horak et al., 1984; Lee et al., 1987; Zattara and Bouisset, 1983), inertial load (Aruin and Latash, 1995b; Benvenuti et al., 1990; Bouisset and Zattara, 1981; Friedli et al., 1984; Kasai and Tanga, 1992; Van der Fits et al., 1998; Zattara and Bouisset, 1988), side (Belenkii et al., 1967; Benvenuti et al., 1990; Bouisset and Zattara, 1981, 1987; Horak et al., 1984; Lee, 1980; Zattara and Bouisset, 1988;), direction (Aruin and Latash, 1995b; Cordo and Nashner, 1982; Friedli et al., 1984; Riach and Hayes, 1990), and orientation in space (Aruin and Latash, 1995a, 1996). In addition, APAs are adaptable; they last longer in the elderly (Horak et al., 1984; Inglin and Woollacott, 1988; Maki, 1993; Stelmach et al., 1990), vanish when persons stay in bed for some weeks (Gurfinkel and Elner, 1973), last longer but are inconsistent in young children (4-6 years) (Riach and Hayes, 1990), and so on. Figure 3.5 illustrates that APAs are graded as a function of movement peak velocity for two inertial load conditions and suggests that both the velocity and the inertial load can affect APAs independently.

Anticipatory postural adjustments are thought to be dynamic, polarized, and task specific, based on the previously mentioned data. They are programmed in relation not only to the focal movement parameters per se but also to the focal movement's location with respect to the body's axes of symmetry.

Role of APAs

The role played by APAs has been argued on the basis of a simple biomechanical analysis (figure 3.3); they tend to create inertial forces that, when the time comes, will counterbalance the disturbance due to the forthcoming intentional movement (Bouisset and Zattara, 1981). Thus, the intentional movement is considered a perturbation, in accordance with the ideas put forward by several neurologists (e.g., André-Thomas, 1940). They were first revisited and checked experimentally by Belenkii et al. (1967), who suggested that the anticipation is related to the constraint of "minimizing the perturbation of the whole kinematic chain." The perturbation applied to the body integrates the effect of movement parameters to the body's structure; in upper limb movement, for example, it is a function of the biacromial width. It can be described by the six components of resultant force and moment. It has been termed *dynamic asymmetry* (Bouisset and Zattara, 1983) and allows a more comprehensive approach to APAs.

In the present context of voluntary movements that do not imply body progression, APAs represent the counterperturbation, which could be supposed either to fix the center of gravity or the joint(s) about which the focal limb is moved, or, for limb movements at least, the reference axes needed to calibrate the motor program, such as the trunk (Gurfinkel et al., 1981). Even if the problem is obviously complex and still under discussion (e.g., Paillard, 1991), the first hypothesis seems to have been largely accepted for the past two decades (see the reviews by Bouisset, 1991, and

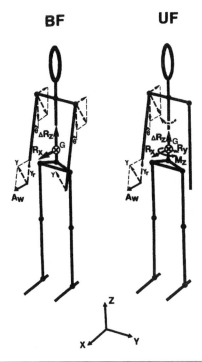

Figure 3.3   Interpretation of the finality of anticipatory postural adjustments. The filled arrows correspond to the actual recorded biomechanical data; the interrupted arrows correspond to theoretical parameters. θ: angular displacement of the upper limb(s); A_w, γ_r, and γ: tangential, radial, and total upper limb accelerations. R_x, R_y, and ΔR_z: anteroposterior, lateral, and vertical accelerations of the body center of gravity, G. M_z: resultant moment about the vertical axis crossing G. BF: bilateral (shoulder) flexion; UF: unilateral (shoulder) flexion.
Reprinted from Bouisset and Zattara 1987.

Massion, 1992).Whether this is totally convincing, for upper limb movements at least, insofar as the resulting perturbation is not very important (Ramos and Stark, 1990); however, the other two possibilities, which are not mutually exclusive, still need to be argued. Nevertheless, as APAs are triggered by a feedforward command and are specific to the motor task characteristics, they have to be determined from previous knowledge of its perturbing effect (Zattara and Bouisset, 1986): anticipatory postural adjustments are programmed according to the expected perturbation rather than to the actual one (Toussaint et al., 1998).

The contention that APAs represent a counterperturbation has been argued within the framework of segmental movements performed on a continuous support base. However, the support base can also change transiently, as in lower limb movements, or periodically, as in locomotion. These movements sill provoke a perturbation of the whole kinematic chain, but the corresponding APAs have been shown to play a more complex role.

In lower limb movements involving a transient change of the support base perimeter, APAs have been shown to be related not only to the counterperturbation to the forthcoming movement but also to the body weight transfer onto the forthcoming stance foot (Do et al., 1991). In locomotion, there is a periodical support base transfer corresponding to a succession of balance losses and recoveries. As has been demonstrated, the faster the progression velocity, the longer the APAs' duration and the greater the forward fall of the COG (Brenière et al., 1987); the APAs induce the postural destabilization that is necessary to initiate the gait. This phenomenon corresponds to an optimal control process, as the propulsive forces at heel-off originate principally from gravity forces, which results in muscular energy saving. In addition, the initiation phase includes a body weight transfer toward the forthcoming stance foot, which can be interpreted as being related to postural stabilization. It is interesting to note that local anticipatory movements have been proved to contribute to one or the other of these global roles; for example, accelerations are occurring at the ipsilateral hip at heel-off, the direction of which suggests that they may oppose the hip perturbation at this instant (Dietrich et al., 1994).

Finally, there are more complex motor tasks that associate various subgoals, as in the one-man-band paradigm, where the music player has to move one or the other upper limb, one or the other lower limb, to step, to turn the head, and so on, in sequence or in phase. The task can include different kinds of conflicting requirements relative to the forthcoming movement parameters, as well as to the postural factors, which have to be simultaneously fulfilled. As the APAs' characteristics are the result of these different requirements, it cannot be excluded that their role necessitates careful interpretation of the analyses.

In conclusion, the APAs appear to play a double role, which consists of postural stabilization and creation of propulsive forces. The proportion of these two constraints appears to vary according to the motor task. In segmental movements performed on a single support base, APAs represent a counterperturbation, which contributes to postural stabilization.

Posturo-Kinetic Capacity As a Limiting Factor of Performance

APAs have been presented as the optimal biomechanical means of initiating voluntary movement and approaching postural programming. To present a more synthetic view, it was tempting to define a posturo-kinetic capacity, as proposed by Bouisset and Zattara (1983): PKC here is the ability to develop a counterperturbation to the postural perturbation induced by segmental movement, and therefore to limit its negative effects on body (or body parts) stability. More generally, PKC can be considered the ability to manage the perturbation to balance associated with the forthcoming movement.

PKC refers to a given motor act. It is assumed to depend particularly on the actual state of the functional systems involved in the intended motor task; for a given subject, it can depend on the postural conditions. More precisely, PKC must be seen as a systemic entity; its properties proceed from a combination of its functional components, mainly the musculoarticular, sensorimotor, and visuomotor systems. It is assumed to be a dynamic process that varies as a function of the instantaneous context, as well as of impairment or rehabilitation, conditioning or fatigue, and development or aging.

To assess PKC, APAs are no longer considered in and of themselves but are referred to the task performance; the corresponding muscular synergy must also be taken into account. PKC assessment can help define the postural strain related to a given task, or to a given subject, in comparison to a reference.

Posturo-Kinetic Capacity, APAs, and Performance

To determine how to evaluate PKC, it is convenient to look briefly at the relation between PKC and performance (figure 3.4). A voluntary movement is triggered consecutively to the intent of performing a given motor task (to reach a target, to manipulate an object, etc.). The intended task, which is the task to perform, can be characterized by different parameters, such as its velocity, amplitude, direction, orientation, or precision. The real task, or the task that is actually performed, fulfills more or less efficiently the intended task requirements. Consequently, PKC, which is the postural counterperturbation, appears to be more or less appropriate for counteracting the task-movement destabilizing effect. In other words, the output of the motor system appears to be a function of the postural input.

Hence, the method for assessing PKC consists of evaluating the postural input to the motor system (the APAs) and the corresponding output (the performance, which is the maximal movement velocity [MMV] or the maximal voluntary force [MVF], etc., according to the paradigm). To this end, anticipatory electromyographic (EMG) activities, as well as anticipatory kinematics and kinetics, can be considered. In addition, the EMG profiles have to be examined and their main features (amplitude, duration, time to peaks, etc.) measured so as to characterize the postural synergy. The influence on PKC of a given factor, such as postural stability or impairment, is assessed in comparing the data to the controls (for example, the results obtained for comfortable stance or able-bodied subjects).

This method can be illustrated with the help of figure 3.5. It appears that the peak velocity of the focal movement is continuously increasing as a function of APAs, up to an APA's threshold. Within the PKC context, figure 3.5 can be said to yield, for a given inertial load, two maximal values: the APAs one defines as the *PKC limit*, and the movement maximal velocity one defines here as the limit performance. In addition, the slope of the relationship between the peak velocity and the APAs is proposed to define a *PKC gain*. When an inertial load is added to the upper limb, the MMV is lessened and the PKC gain is lower; a greater PKC percentage is invested into the motor act for a minor performance. In addition, the postural synergy is the same in both conditions (Horak et al., 1984; Lee et al., 1987); there are only modifications in the muscle's excitation level, timing, and synergist recruitment.

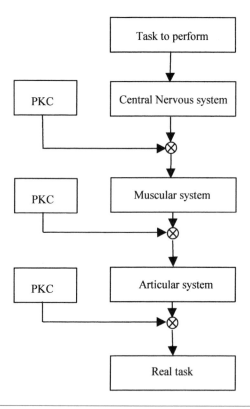

Figure 3.4   Task to perform, real task, and posturo-kinetic capacity (PKC): outline of a system analysis. The overall command process of an efficient motor act can be conceived as hierarchical and including three functional levels: planning, programming, and execution. All information concerning the parameters of different orders that characterize the task to perform is sent to the programming level. A program is then selected to fit the task parameters and the initial conditions. The program issues a central command that triggers signal commands to the muscular system, which controls the focal and the postural chains. At each instant, the real task is the outcome of two opposite effects: the perturbation and the counterperturbation. PKC is defined by the functional state of the composing systems, which are schematized as central nervous, muscular, and articular systems. If the demands exceed the PKC limit at one level or another, the parameters of the real task are different from those of the intended task: the motor activity is less efficient.

A simpler procedure for assessing PKC consists of considering the task movement performed at maximal velocity. If the MMV is slower than in the controls and the APAs last longer, their amplitude being smaller, as in paraplegics, it is concluded that PKC is reduced. In addition, if the postural synergy is different, a less economical process is assumed to have been used, insofar as the primary synergy can be supposed to result from an optimization criterion: there is an increased *PKC cost*.

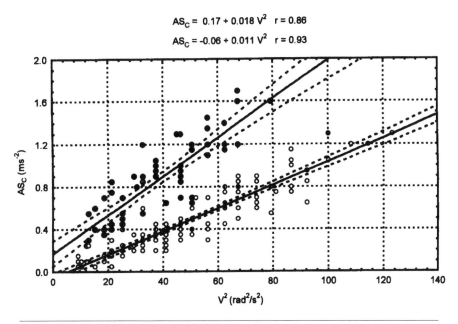

$AS_C = 0.17 + 0.018 \, V^2 \quad r = 0.86$

$AS_C = -0.06 + 0.011 \, V^2 \quad r = 0.93$

Figure 3.5   Relationships between anticipatory peak acceleration amplitudes at the contralateral shank and squared peak velocity of the forthcoming movement. APA scales are in ms^{-2}; V^2 in rad^2/s^2. Regression line equations and correlation coefficients are presented for the conditions IUF (upper solid line) and OUF (lower solid line). Data for movements without additional inertia (OUF) are shown in open circles; movements with additional inertia (IUF) are shown with filled circles. The 95% confidence limits for the regression lines are shown with dashed lines.

Reprinted from Bouisset, Richardson, and Zattara 2000.

Posturo-Kinetic Capacity and Body Balance Stability

As discussed previously, the "static" body postures, which precede and follow the voluntary movement, correspond to stable states. They can be evaluated by the stability indexes. Voluntary movement takes place between these two stable states; it transiently perturbs the initial one and makes the final one more or less difficult to stabilize. Insofar as the support reaction forces (and moments) are known to determine body dynamics, PKC likely depends on postural conditions.

Only a few studies have been devoted to the question, however. They focused on the relation among postural stability, APAs, and/or muscular synergy, but the performance was rarely considered. Various task-movements have been studied, such as moving the upper limb (Cordo and Nashner, 1982), moving the lower limb (Rogers and Pai, 1990), or rising onto tiptoes (Lipshits et al., 1981) and rocking on the heels (Nardonne and Schieppati, 1988). They differ as to the body segments that are voluntarily moved; they also differ as to the support base conditions they induce, which may be continuous or transient according to the paradigm. As they exercise numerous influences on APAs, the dependence of APAs on the stability demands is complex to analyze and may have appeared somewhat difficult to interpret.

Two main experimental situations have been considered. In the first, the voluntary movement was performed while the support base was oscillating, whether the oscillations originated from the proper instability of the support on which the subject remains while the movement is performed or in the external perturbations applied to it.

For some authors, when stability was low, the anticipatory postural EMGs tended to increase. Thus, Cordo and Nashner (1982), in their study on pulling on a handle, reported that APAs were reduced when the subject was given an additional thoracic support; in contrast, APAs were increased when the support surface was continuously oscillating. According to Gantchev and Dimitrova (1996), the anticipatory EMGs of postural muscles, associated with upper limb elevations, started earlier during balancing on an unstable support surface (seesaw), whereas their amplitude was smaller.

In contrast, Pedotti et al. (1989) found that the anticipatory EMGs were reduced when the subjects performing trunk extensions were standing on a small plate placed on a narrow support and thus were more oscillating. Their data were confirmed by Aruin et al. (1998); according to these authors, standing on a platform with a narrow support area led to an attenuation of anticipatory EMGs associated with horizontal upper limb abductions, and the effects were stronger when the initial instability was in the sagittal rather than in the frontal plane. Unfortunately, the MMV was not measured in these studies.

To explain the discrepancies between the results, it is helpful to refer to Nouillot et al. (1992), who reported that APAs were absent when subjects were standing on one foot from the beginning to the end of a lower limb movement; as APAs induce dynamics, they can potentially perturb body balance, and they are no longer present when the posture is too unstable. Therefore, in conditions of high stability demand, the CNS may reduce and even suppress APAs as a protection against their possible destabilizing effects. The focal and the postural chains are then activated simultaneously at the expense of greater forces to be developed during the movement itself.

In the second experimental situation, the voluntary movement was performed on a reduced support base configuration, which included the perimeter and the body parts in contact with the physical support. For example, APAs duration and MMV have been reported (Zattara and Bouisset, 1994) to vary in opposite directions when the support base was reduced; when two-legged stance position was modified, from the feet normally spread (FNS) to the feet close together (FCT), the general trend was an increase in APAs duration, whereas MMV decreased. Thus, despite an increase in anticipatory postural dynamics, the maximal movement velocity is slower when the support base perimeter is reduced. In addition, the timing and excitation level of postural EMGs depend mainly on the forthcoming movement characteristics, but the postural synergy does not change when the support base configuration is changed.

Most of the studies related to the influence of body posture have been done on lower limb movements, which perturb body balance stability for two reasons: due to the movements themselves and to the transient changes of body contact areas they induce. They have been developed by various authors who have studied the influence on APAs of the initial and final postures (Aruin et al., 1998; Béraud and Gahéry, 1997; Do et al., 1991; Mouchnino et al., 1991; Nouillot et al., 1992; Rogers, 1992). The results established that APAs amplitude and duration are functions of both the initial and final postures and suggest that they also depend on movement dynamics. They are in agreement with the pioneering study on rising onto tiptoes of Lipshits et al. (1981), whose results were confirmed by Nardonne and Schieppati (1988) in a study on rocking on the heels. In addition, Lipshits et al. (1981) have shown that APAs progressively decrease and even disappear when the trunk is progressively inclined forward; they are no longer observed if the subject comes back to the initial posture instead of staying on tiptoes at the end of the forward movement.

In summary, these results demonstrate that APAs (and PKC) are related to body balance stability insofar as they are known to depend on instantaneous physical support instability and body posture, particularly support base configuration. Using more biomechanical terms, APAs can be said to depend on dynamic as well as on static equilibrium conditions. Therefore, postural stability is supposed to be a parameter of motor programming. The influence on APAs of the focal movement and postural parameters is cumulative insofar as they both contribute to postural destabilization. The added requirements they induce result in increased APAs when postural instability is increased (at least up to a certain limit), the maximal performance usually being reduced (and at best not different). As the APAs are lengthened, the PKC gain, and hence PKC, are reduced.

In addition, in most of the studies under consideration, the general pattern of postural reactions, including anticipatory postural adjustments, was preserved regardless of the increased requirements for maintaining body balance stability. In other words, the postural synergy associated with upper limb movement remained unchanged, and the adaptations were limited to changes in EMG amplitude, timing, and the like. The change from the primary synergy to another occurs only when the body stability is too precarious (Nouillot et al., 1992). If the primary synergy is supposed to be the optimal one, the new one corresponds to a less economical process, which results in increased dynamics after the onset of the focal movement. In this case, PKC is not necessarily lessened, but the PKC cost is very likely greater: the postural synergy complies with the efficiency rather than the economy principle.

Posturo-Kinetic Capacity and Mobility of the Postural Chain

As stated in the preceding sections, the voluntary movement is a perturbation to body balance, and the postural counterperturbation must be adequate for the intended performance to be achieved. Therefore, if the movement induces a dynamic perturbation, the counterperturbation must be dynamic as well. Consequently, if the postural chain mobility is constrained in any way, fewer postural segments could be accelerated, the counterperturbation would be limited, and the performance reduced.

An interesting way to examine this possibility in more detail was to consider a new paradigm with upper limb tasks performed in two sitting postures that differ only in the percentage of ischiofemoral contact with the seat (100% or 30%). In these conditions, the support base perimeter remains the same, but the pelvis mobility is reduced when the ischiofemoral contact is increased.

Diverse experimental series have been undertaken on subjects performing a pointing task. Lino and Bouisset (1994) have shown that a reduction of the ischiofemoral contact area elicits MMV as well as APAs increases. These data have been confirmed in isometric ramp efforts performed as rapidly as possible (Le Bozec et al., 1997, 1998); they display dynamic postural adjustments, and the greater these adjustments, the longer the APAs and the larger the maximal isometric force. Thus, a contact area reduction leads to increased performance in "static" (but "anisotonic") as well as "dynamic" conditions (figure 3.6). These observations are interpreted as supporting the view that any variation of the external force perturbs the subject's balance and that the maximal value of this force depends on the intensity of the dynamic postural counterperturbation. These results agree with those of Van der Fits et al. (1998), who reported that larger support surfaces induce an MMV limitation, MMV decreasing stepwise from upright sitting to long-leg sitting and semireclined sitting.

To delve deeper into the influence of the support contact area, the results of a pointing task performed in two standing postures were examined and compared to those obtained in two sitting postures (Goutal et al., 1994). The subjects were standing with their feet normally spread or close together and sitting with a full (100 IF) or a one-third (30 IF) ischiofemoral contact, as previously. The main results showed that the MMV was significantly faster (and APAs significantly longer) in FNS than in 100 IF conditions (table 3.1) (i.e., when the support base perimeter and hence the stability were reduced). In addition, the MMV was confirmed to be significantly slower (and the APAs shorter) for FCT than for FNS, which was interpreted earlier as resulting from reduction of the support base perimeter. In the present context, one could ask whether placing the feet, and therefore the lower limbs, close together reduces the postural chain mobility as well as the support base perimeter.

In addition, as in the quiet posture maintenance, the general pattern of postural reactions, including APAs, depends on the category of body posture (standing, sitting, etc.). For a given posture category, the pattern was preserved regardless of the requirements related to the mobility of the postural chain. In other words, the postural synergy associated with upper limb movements remained unchanged, and the adaptations were limited to changes in EMG amplitude, timing, and the like. Consequently, in accordance with Latash and Anson (1996), it can be claimed that posture control consists mainly of adapting the motor program according to the postural requirements rather than consisting of changes in the postural strategy.

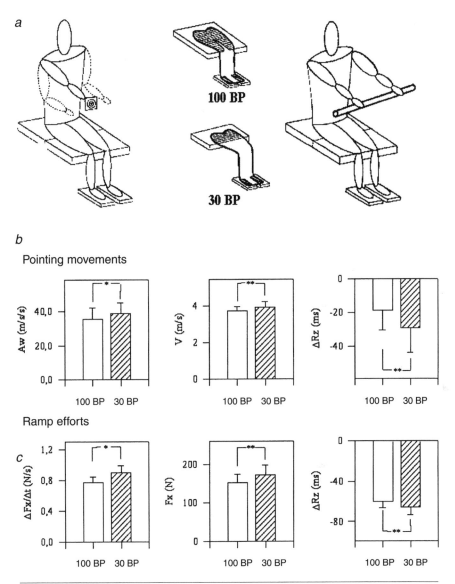

Figure 3.6 APAs and performance for pointing movements and ramp isometric efforts. *(a)* Shows the two experimental paradigms and the ischiofemoral contact conditions. 100 BP (bilateral push): full ischiofemoral contact; 30 BP: 30% ischiofemoral contact. *(b)* Pointing movements, from left to right: Aw and V are the wrist peak acceleration and velocity; ΔRz is the APAs measured on the vertical reaction force component (Lino and Bouisset, 1994). *(c)* Ramp efforts, from left to right: ΔFx/Δt and Fx are the peak isometric force and its first derivate; ΔRz is the APAs measured on the vertical reaction force component (Le Bozec et al., 1998).

*p < .05, ** p < .01.

Table 3.1   APAs for the Same Pointing Movement Performed in Four Different Postures

		Rx (ms)		ΔRz (ms)		V (m/s)	
		m ± s	p	m ± s	p	m ± s	p
Sta	FNS	−20 ± 6		−28 ± 10		3.99 ± 0.25	
	FCT	−14 ± 5		−21 ± 5		3.81 ± 0.21	
Sit	100 IF	−3 ± 25		−19 ± 14		3.73 ± 0.27	
	30 IF	−13 ± 20		−29 ± 18		3.93 ± 0.31	

The APAs were measured along the vertical (ΔRz) reaction forces component; V is the peak velocity. The subjects were standing (Sta) with their feet normally spread (FNS) or close together (FCT), and sitting (Sit) with a full (100 IF) or a one-third (30 IF) ischio-femoral contact. Means (m) ± one standard deviation (s); p is the probability.

*p < .05

**p < .01

In conclusion, postural chain mobility, as well as body balance stability, appears to be a postural parameter. Both should be taken into account in motor programming. Even if it can be supposed that the former rules over the latter, they both contribute to PKC.

Posturo-Kinetic Capacity and Impairments

Within the present conceptual context, if PKC is reduced, a given movement velocity would induce a relatively higher perturbation and APAs have to be more important. However, APAs by themselves constitute a perturbation to postural balance (Bouisset, 1991; Nouillot et al., 1992). Therefore, it could be assumed that APAs must be limited in proportion to the lessening of postural stability or mobility, and that longer APAs durations are more convenient than higher APAs amplitudes. Such a hypothesis allows the interpretation of different results obtained in sensorimotor impairments.

In paraplegics performing a pick-and-place task, Do et al. (1985) have established that the MMV is slower than in healthy controls but that the APAs last longer and their amplitude is reduced. In addition, the postural synergy is different; other scapular girdle and trunk muscles are activated, especially on the contralateral side (figure 3.7). Assuming that the primary synergy corresponds to an optimization

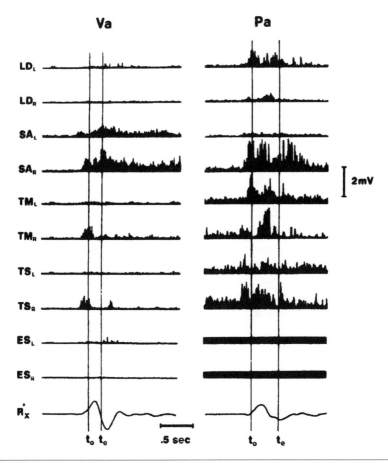

Figure 3.7   EMG pattern in paraplegic (Pa) compared to able-bodied (Va) subjects performing a pick-and-place task. From top to bottom: latissimus dorsi (LD), serratus anterior (SA), trapezius medialis (TM), trapezius superior (TS), erectores spinae (ES), which cannot be activated by paraplegics. The subscripts R and L indicate right and left, respectively. R_x^* is the anteroposterior component of the normalized resultant of forces; t_o and t_e are the onset and the end of the forward load transport. Superimposition of five records of rectified EMGs.

From Do, Bouisset, and Moynot 1985.

criterion, the transition from one synergy to another can be supposed to correspond to a less economical process. Thus, despite a change in the postural synergy, the PKC limit and the PKC gain are decreased, whereas the PKC cost is increased. In other words, paraplegics, whose equilibrium is more unstable than able-bodied subjects, could not develop, without an excessive balance risk, postural adjustments adapted to the perturbation provoked by the motor task; the perturbation should be reduced, and hence the required performance. Therefore, voluntary movement is less efficient, thanks to PKC reduction.

In subjects with Parkinson's disease, the main results were the change in postural strategy and the lack of APAs associated with bradykinesia (Bazalgette et al., 1986; Bouisset and Zattara, 1990). It may raise the question: Was the bradykinesia in Parkinson's disease due to pathological postural adjustments, or did the pathological postural adjustments and bradykinesia originate from the same neurological impairment? Some of the previous data are not in disagreement with the first hypothesis. Indeed, for submaximal movements, APAs have not been observed under 4 rad/s in healthy subjects (Zattara and Bouisset, 1983), and this value could correspond to a threshold below which the inertial forces due to the intentional movement become negligible. The upper limit for the movement velocity in Parkinson's patients, which is about the same value, could be explained by their inability to counterbalance these perturbing forces. Moreover, it is interesting to note that this limit is higher when the subjects are seated, in contrast to able-bodied subjects.

Therefore, the results obtained in paraplegics as well as in Parkinson's patients could be interpreted in terms of PKC. The same hypothesis also fits diverse results such as those obtained in hemiplegics (Horak et al., 1984) and the elderly (Stelmach et al., 1990), according to which the maximal movement velocity is reduced, whereas APAs are longer. Moreover, the postural patterns are observed to change in paraplegics, Parkinson's patients, hemiplegics, and so forth. The problem of adaptive change in motor patterns of "atypical" individuals has been discussed in detail by Latash and Anson (1996).

In conclusion, the ability to develop postural regulations in relation to the perturbation induced by intentional movement, or PKC, is lessened in sensorimotor impairments due to a loss, a lessening, or a dysfunction of pelvis and lower limb motor control. This statement holds true for all traumatic or pathological impairments that result in a restriction of joint mobility, whatever its exact origin (articular, tendinous, muscular, or nervous).

Postural Chain Dynamic Mobility and Body Balance Stability

The preceding results have established that PKC depends not only on body balance stability but also on another postural factor: postural chain mobility. The biomechanical content of the latter is classical, contrary to that of the former. Indeed, postural chain mobility has been presented as the ease with which postural segments are moved or, more precisely, accelerated; it is a *dynamic mobility* that is considered. The problem is to examine the biomechanical role of this dynamic mobility and to precisely state its relation to body balance associated with voluntary movement.

Human body balance stability, as stated earlier, must comply with global and local requirements. The global requirements result in the contention that general body balance is secured when the gravity and the support reaction forces are equal and opposite in sign; in other words, the projection on the ground of the center of gravity falls at the center of pressure (and hence within the support base perimeter).

The data on the influence on PKC of postural oscillations and support base configuration have been explained in terms of preservation of the global body balance. The local requirements result from the fact that each body part must also be balanced with respect to the underlying one. We propose that postural chain mobility is related to this second set of requirements for reasons that are delineated in the following section.

The postural chain dynamic mobility results from anatomical and physiological factors. From an anatomical viewpoint, the mobility of an articular chain is a function of the range of individual joint movements, which results from their anatomical structures (bony stops, ligaments, tissue bulks, etc.); they display measurable differences originating from such factors as age, training, and disease. However, data on isolated joints cannot entirely reflect the cumulative geometric background for the postural link system, and it is also important to know the relative orientation of one joint to the next in sequence. In other words, the mobility of an articular chain should be modified in relation to the body posture; it is, a fortiori, the same when a physical support such as a seat is used, which limits the range of the joint's movement (figure 3.8). From a physiological viewpoint, the dynamic mobility of an articular chain is a function of the muscular properties, which, for a given excitation pattern, result in muscular tension (and stiffness). The body segment acceleration originates from the muscular moments acting on the joints. For a given joint, the muscular moment varies as a function of the joint angle.

Therefore, if body posture or impairment induces smaller joint amplitude ranges or lesser muscular moments, the acceleration of one or the other postural segment can be limited and the counterperturbation eventually reduced. Consequently, PKC

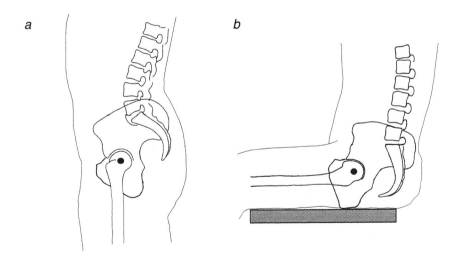

Figure 3.8   Mobility of the pelvis according to the posture. *(a)* Standing posture; *(b)* sitting posture. The change of posture induces a 90° reduction of hip flexion. In addition, the mobility of the lumbar column is modified, given that it moves from lordosis to kyphosis.

could also be reduced, except for a muscle excitation increase or a postural strategy change.

There was, a priori, no reason to conclude that these considerations are no longer valid when balance is perturbed by external forces. Indeed, similar considerations were applied successfully by Roberts (1978) in his analysis on the effect on postural stabilization of restricted perimeter area and the return from the brink. The COG can be maintained within the stability cone, thanks to the opposite directions of adjacent body link movements, which provide partitioning of the angular momentum of the body; each segmental counterperturbation results from the inertial forces, the intensity of which is, for a body segment, a function of its inertia and angular acceleration. Consequently, as the intensity of the acceleration depends on the joint displacement amplitude, the performance (the balance recovery) is a function of the postural chain mobility (figure 3.9). If necessary, one compensatory strategy can be replaced by another (e.g., the ankle strategy by the hip strategy), as in the Horak and Nashner (1986) experiments. If a compensatory strategy is not sufficient, the CNS institutes corrective movements, where the subject has to step to recover balance.

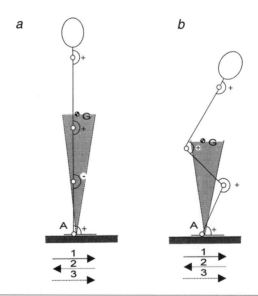

Figure 3.9 Postural chain dynamic mobility and body balance recovery. In response to a support translation directed forward (indicated by arrow 1), there is a backward passive body movement (2). Contractions of the anterior plane muscles (3) are necessary for balance recovery. The accelerations of postural body segments are a function of the joint displacement and muscular moments. The intensity of muscular moments and the muscles that are activated (+) depend on the initial posture, which constrains the joint's mobility. For example, *(a)* in the upright standing posture, there is no possibility of extending the knee (–), *(b)* contrary to the crouch posture. Note that the stability cone is not symmetrical with regard to the vertical insofar as the ankle joint (A) is not located in the middle of the foot. G is the body's center of gravity.

Conclusion: Posturo-Kinetic Capacity, Dynamic Body Structure, and Postural Function

The PKC approach was adopted in this review to probe deeper into the postural function in voluntary movement. It entailed posing the problem in terms of focal and postural chains. Within this conceptual frame, voluntary movement, which is the movement of the focal chain, constitutes a perturbation to body balance (i.e., to the postural chain). But if the perturbation originates from the focal chain, it is applied to the body, and its effects depend on its location on the body; in other words, it results from integration of the effect of movement parameters with the body's anatomical structure (dynamic asymmetry). Moreover, as the perturbation is a dynamic process, the counterperturbation must also be dynamic.

PKC has been defined as the ability to counteract the perturbation induced by the focal movement. It has been shown to depend not only on body balance stability but also on another postural factor: the postural chain dynamic mobility. Neither of these can be considered apart from the body structure characteristics. Indeed, body balance stability and postural chain dynamic mobility are functions of the range of the postural joint movements, which is based on anatomical structures (bony stops, ligaments, tissue bulks, etc.), the configuration of body segments (i.e., body posture), and subject's morphology.

Anatomical characteristics therefore constitute a part of the postural system, which it is not unusual to disregard. They must be taken into account in the postural component of the motor program, but they do not interfere with the postural process as "static" structures; the postural as well as the focal chains are dynamic structures and must be controlled as such. Thus, PKC can be considered a dynamic link between anatomical body structure and postural function.

References

Allum, J.H.J. (1990) Posturography systems: Current measurement concepts and possible improvements. In Brandt, T., et al., *Disorders of posture and gait*, pp. 16-28. Stuttgart: G. Thieme Verlag.

André-Thomas (1940) *Equilibre et équilibration*. Paris: Masson.

Aristotle (1973) *Animal locomotion*. Paris: Société "Les Belles Lettres."

Aruin, A.S., Forrest, W.R., and Latash, M.L. (1998) Anticipatory postural adjustments in conditions of postural instability. *Electroenceph Clin Neurophysiol* 109: 350-359.

Aruin, A.S., and Latash, M.L. (1995a) The role of motor action in anticipatory postural adjustments studied with self-induced and externally triggered perturbations. *Exp Brain Res* 106: 291-300.

Aruin, A.S., and Latash, M.L. (1995b) Directional specificity of postural muscles in feed-forward postural reactions during fast voluntary arm movements. *Exp Brain Res* 103: 323-332.

Aruin, A.S., and Latash, M.L. (1996) Anticipatory postural adjustments during self-initiated perturbations of different magnitude triggered by a standard motor action. *Electroenceph Clin Neurophysiol* 101: 497-503.

Babinski, J. (1899) De l'asynergie cerebelleuse. *Rev Neurol* 7: 806-816.

Bazalgette, D., Zattara, M., Bathien, N., Bouisset, S., and Rondot, P. (1986) Postural adjustments associated with rapid voluntary arm movements in patients with Parkinson's disease. In Yarr, P.D., and Bergman, K.J., *Advances in Neurology* 45 (Raven Press): 371-374.

Belenkii, Y.Y., Gurfinkel, V.S., and Paltsev, Y.I. (1967) Element of control of voluntary movements. *Biofizika* 12: 135-141.

Benvenuti, F., Panzer, V., Thomas, S., and Hallett, M. (1990) Kinematic and EMG analysis of postural adjustments associated with fast elbow flexion movements. In Brandt, T., et al., *Disorders of posture and gait,* pp. 72-75. Stuttgart: G. Thieme Verlag.

Bernstein, N. (1935) *Coordination and regulation of movements.* Elmsford, NY: Pergamon Press (Amer. translation, 1967).

Béraud, P., and Gahéry, Y. (1997) Posturo-kinetic effects on kicking movements of a lack of initial ground support under the moving leg. *Neurosci Lett* 226: 5-8.

Bouisset, S. (1991) Relation entre support postural et mouvement intentionnel: Approche biomécanique. *Arch Int Physiol Bioch Biophys* 99: A77-A92.

Bouisset, S., and Maton, B. (1995) Muscles, posture et mouvements. Paris: Hermann éd.

Bouisset, S., Richardson, J., and Zattara, M. (2000) Do APAs occurring in different segments of the postural chain follow the same organisation rule for different task movement velocities, independently of the inertial load value? *Exp Brain Res* 132: 79-86.

Bouisset, S., and Zattara, M. (1981) A sequence of postural movements precedes voluntary movement. *Neurosci Lett* 22: 263-270.

Bouisset, S., and Zattara, M. (1983) Anticipatory postural movements related to a voluntary movement. In *Space physiology,* pp. 137-141. Toulouse: Cepadues Editions.

Bouisset, S., and Zattara, M. (1987) Biomechanical study of the programming of anticipatory postural adjustments associated with voluntary movement. *J Biomech* 20(8): 735-742.

Bouisset, S., and Zattara, M. (1990) Segmental movement as a perturbation to balance? Facts and concepts. In Winters, J.M., and Woo, S.L.-Y. (Eds.), *Multiple muscle system, biomechanics and movement organization,* pp. 498-506. New York City: Springer Verlag.

Brenière, Y., Do, M.C., and Bouisset, S. (1987) Are dynamic phenomena prior to stepping essential to walking? *J Motor Behav* 19(1): 62-76.

Brown, J.E., and Frank, F.S. (1987) Influence of event anticipation on postural actions accompanying voluntary movement. *Exp Brain Res* 67: 645-650.

Clément, G., Gurfinkel, U.S., Lestienne, F., Lipshits, M.I., and Popov, K.E. (1984) Adaptation of postural control of weightlessness. *Exp Brain Res* 57: 61-72.

Cordo, P.J., and Nashner, L.M. (1982) Properties of postural movements related to a voluntary movement. *J Neurophysiol* 47: 287-303.

Crenna, P., Frigo, C., Massion, J., and Pedotti, A. (1987) Forward and backward axial synergies in man. *Exp Brain Res* 65: 538-548.

Dempster, W.T. (1955) The anthropometry of body action. *Annals N.Y. Academy of Sciences* 63: 559-585.

Dick, J.P.R., Rothwell, J.C., Berardelli, A., Thompson, P.D., Gioux, M., Benecke, R., Day, B.L., and Marsden, C.D. (1986) Associated postural adjustments in Parkinson's disease. *J Neurol Neurosurg Psychiat* 49: 1378-1385.

Dietrich, G., Brenière, Y., and Do, M.C. (1994) Local expressions of anticipatory movements in single step initiation. *Human Movement Studies* 13: 195-210.

Do, M.C., Bouisset, S., and Moynot, C. (1985) Are paraplegics handicapped in the execution of a manual task? *Ergonomics* 28(9): 1363-1375.

Do, M.C., Nouillot, P., and Bouisset, S. (1991) Is balance or posture at the end of a voluntary movement programmed? *Neurosci Lett* 130: 9-11.

Friedli, W.G., Cohen, L., Hallett, M., Stanches, S., and Simon, S.R. (1988) Postural adjustments associated with rapid voluntary arm movements: II. Biomechanical analysis. *J Neurol Neurosurg Psychiat* 51: 232-243.

Friedli, W.G., Hallett, M., and Simon, S.R. (1984) Postural adjustments associated with rapid voluntary arm movements: I. Electro-myographic data. *J Neurol Neurosurg Psychiat* 47: 611-622.

Gantchev, G.N., and Dimitrova, D.M. (1996) Anticipatory postural adjustments associated with arm movements during balancing on unstable support surface. *Int J Psychophysiol* 22: 117-122.

Goutal, L., Lino, F., and Bouisset, S. (1994) Modalités de l'appui corporel et vitesse du mouvement de pointage. *Arch Int Physiol Bioch Biophys* 102(5): C21.

Gray, J. (1968) *Animal locomotion.* London: Weidenfeld and Nicolson.

Gurfinkel, V.S., and Elner, A.M. (1973) On two types of static disturbances in patients with local lesions of the brain. *Agressol* 14D: 65-72.

Gurfinkel, V.S., Lipshits, M.I., Mori, S., and Popov, K.E. (1981) Stabilization of body position as the main task of postural regulation. *Fiziol Cheloveka* 7: 400-410.

Hellebrandt, F.A., Houtz, S.J., Partridge, M.J., and Walters, C.E. (1956) Tonic reflexes in exercises of stress in man. *Amer J Phy Med* 35: 144-159.

Horak, F.B., Esselman, P.E., Anderson, M.E., and Lynch, M.K. (1984) The effect of movement velocity, mass displaced and task certainty on associated postural adjustments made by normal and hemiplegic individuals. *J Neurol Neurosurg Psychiat* 47: 1020-1028.

Horak, F.B., and Nashner, L.M. (1986) Central programming of postural movements: Adaptation to altered support-surface configuration. *J Neurophysiol* 55(6): 1369-1381.

Houtz, S.J. (1964) Influence of gravitational forces on function of lower extremity muscles. *J Appl Physiol* 19: 999-1004.

Houtz, S.J., and Fisher, F.J. (1961) Function of leg muscles acting on foot as modified by body movements. *J Appl Physiol* 16: 597-605.

Inglin, B., and Woollacott, M. (1988) Age-related changes in anticipatory postural adjustments associated with arm movements. *J Gerontology* 43: M105-113.

Kasai, T., and Tanga, T. (1992) Effects of varying load conditions on the organisation of postural adjustments during voluntary flexion. *J Motor Behav* 24: 359-365.

Latash, M.L., and Anson, J.G. (1996) What are "normal movements" in atypical populations? *Behav Brain Sci* 19: 55-106.

Le Bozec, S., Goutal, L., and Bouisset, S. (1997) Dynamic postural adjustments associated with the development of isometric forces in sitting subjects. *C R Acad Sci Paris* 320: 715-720.

Le Bozec, S., Lesne, J., and Bouisset, S. (1998) Postural muscles activation associated with isometric ramp postural muscles. In Arsenault, B., McKinley, P., and McFadyen, B. (Eds.), *Twelfth Congress of the International Society of Electrophysiology and Kinesiology,* pp. 251-252. Montréal: McGill University.

Lee, W.A. (1980) Anticipatory control of posture and task muscles during rapid arm flexion. *J Motor Behav* 12: 185-196.

Lee, W.A., Buchanan, T.S., and Rogers, M.W. (1987) Effects of arm acceleration and behavioral conditions on the organization of postural adjustments during arm flexion. *Exp Brain Res* 66: 257-270.

Lino, F., and Bouisset, S. (1994) Is velocity of a pointing movement performed in a sitting posture increased by upper body instability? *J Biomech* 27: 733.

Lipshits, M., Mauritz, K., and Popov, K.E. (1981) Quantitative analysis of anticipatory postural components of a complex voluntary movement. *Hum Physiol* 7: 411-419.

Maki, B.E. (1993) Biomechanical approach to quantifying anticipatory postural adjustments in the elderly. *Med Biol Eng & Comput* 31: 355-362.

Marey, E.J. (1883) De la mesure dans les différents actes de la locomotion. *C R Acad Sci Paris* 97: 820-825.

Martin, J.P. (1967) *The basal ganglia and posture.* London: Pitman.

Massion, J. (1992) Movement, posture and equilibrium: Interaction and coordination. *J Prog Neurobiol* 35: 35-56.

McCollum, G., and Leen, T.K. (1989) Form and exploration of mechanical stability limits in erect stance. *J Motor Behav* 21: 225-244.

Mouchnino, L., Aurenty, R., Massion, J., and Pedotti, A. (1991) Stratégies de contrôle simultané de l'équilibre et de la position de la tête pendant l'élévation d'une jambe. *C R Acad Sci Paris* 312: 225-232.

Murray, M.P., Seireg, A., and Scholz, R.C. (1967) Center of gravity, center of pressure and supportive forces during human activities. *J Appl Physiol* 23: 831-838.

Murray, M.P., Wood, A.A., Seireg, A., and Sepic, S.B. (1975) Normal postural stability and steadiness: Quantitative assessment. *J Bone Joint Surg* 57A: 510-516.

Nardone, A., and Schieppati, M. (1988) Postural adjustments associated with voluntary contractions of leg muscles in standing man. *Exp Brain Res* 69: 469-480.

Nashner, L.M. (1976) Adapting reflexes controlling human posture. *Exp Brain Res* 26: 59-72.

Nouillot, P., Bouisset, S., and Do, M.C. (1992) Do fast voluntary movements necessitate anticipatory postural adjustments even if equilibrium is unstable? *Neurosci Lett* 147: 1-4.

Oddsson, L., and Thorstensson, A. (1986) Fast voluntary trunk flexion movements in standing: Primary movements and associated postural adjustment. *Acta Physiol Scand* 128: 341-349.

Paillard, J. (1991) *Brain and space.* New York: Oxford University Press.

Pedotti, A., Crenna, P., Deat, A., Frigo, C., and Massion, J. (1989) Postural synergies in axial movements: Short- and long-term adaptation. *Exp Brain Res* 74: 3-10.

Ramos, C.F., and Stark, L.W. (1990) Postural maintenance during movement: Simulations of a two joint model. *Biol Cybern* 63: 363-375.

Riach, C.L., and Hayes, K.C. (1990) Anticipatory postural control in children. *J Motor Behav* 22: 250-256.

Roberts, T.D.M. (1978) *Neurophysiology of postural mechanisms*. London: Butterworths.

Rogers, M.W. (1992) Influence of task dynamics on the organization of interlimb responses accompanying standing human leg flexion movements. *Brain Res* 579: 353-356.

Rogers, M.W., and Pai, Y.C. (1990) Dynamic transitions in stance support accompanying leg flexion movements in man. *Exp Brain Res* 81: 398-402.

Stapley, P., Pozzo, T., and Grishin, A. (1998) The role of anticipatory postural adjustments during whole body reaching movements. *NeuroReport* 9: 395-401.

Stelmach, G.E., Populin, L., and Friedemann, M. (1990) Postural muscle onset and voluntary movement in the elderly. *Neurosci Lett* 117: 188-193.

Toussaint, H.M., Michies, Y.M., Baber, M.N., Commissaris, A.C.M., and van Dieën, J.H. (1998) Scaling anticipatory postural adjustments dependent on confidence of load estimation in a bimanual whole-body lifting task. *Exp Brain Res* 120: 85-94.

Van der Fits, I.B.M., Klip, A.W.J., and Van Eykern, L.A. (1998) Postural adjustments accompanying fast pointing movements in standing, sitting and lying adults. *Exp Brain Res* 120: 202-216.

Vernazza, S., Cincera, M., Pedotti, A., and Massion J. (1996) Balance control during lateral arm raising in humans. *NeuroReport* 7: 1543-1548.

Wing, A.M., Randall, F.J., and Richardson, J. (1997) Anticipatory postural adjustments in stance and grip. *Exp Brain Res* 116: 122-130.

Zattara, M., and Bouisset, S. (1983) Influence de la vitesse d'exécution du mouvement volontaire sur les accélérations locales anticipatrices. *8ème Congrès de la Société de Biomécanique,* Lyon, in abstracts: 113-114.

Zattara, M., and Bouisset, S. (1986) Chronometric analysis of the posturo-kinetic programming of the voluntary movement. *J Motor Behav* 18: 215-223.

Zattara, M., and Bouisset, S. (1988) Posturo-kinetic organization during the early phase of voluntary upper limb movement: 1. Normal subjects. *J Neurol Neurosurg Psychiatry* 51: 956-965.

Zattara, M., and Bouisset, S. (1994) Reduction of postural base configuration lessens the velocity of upper limb movements performed in standing posture. *J Biomech* 27 (6): 748.

CHAPTER

Neuronal Mechanisms Underlying Postural Control As Revealed in Simpler Systems

G.N. Orlovsky and T.G. Deliagina
The Nobel Institute of Neurophysiology,
Department of Neuroscience,
Karolinska Institutet

Y.I. Arshavsky
Institute for Nonlinear Science,
University of California at San Diego;
Institute of Information Transmission Problems,
Russian Academy of Sciences

Different species, from mollusk to man, maintain a particular body orientation in space (the upright body posture) both in a quiescent state and during locomotion. This posture also presents a basis on which voluntary movements of different parts of the body can be superimposed. The upright body posture is maintained due to the activity of the postural control system, a general functional organization for which is shown in figure 4.1 (Ghez, 1991; Horak and Macpherson, 1995; Massion, 1994). The system operates on the basis of information delivered by vestibular, visual, and somatosensory inputs. These signals are processed and integrated. If they show that

Sensory inputs

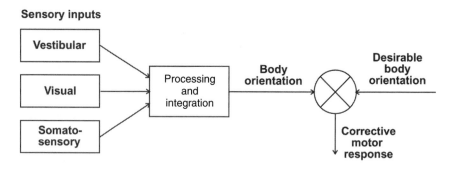

Figure 4.1 General functional organization of the postural control system. Sensory signals are processed and integrated to characterize the body orientation. If the orientation differs from the desirable one, a corrective motor response is generated.

the body orientation differs from the desirable one, a corrective motor response is generated. The response is aimed at restoration of the normal orientation.

Studies in the field of postural control are devoted to different functional aspects of this general scheme. The most common method in these studies is the observation of motor and electromyographic responses to postural disturbances under different conditions. These studies led to formulation of several conclusions and concepts.

- *Processing and integration of sensory inputs.* The relative role of different sensory inputs in postural stabilization was found to be species-dependent. In particular, vestibular input plays a much larger role in aquatic animals than in terrestrial ones (Burt and Flohr, 1991; Magnus, 1924; Orlovsky, 1991). A relative contribution of inputs of different modalities for a particular species is not constant but may vary considerably depending on the behavioral state of the animal and on many other factors, which can be interpreted as a substitution of a given input by other inputs (Horak and Macpherson, 1995; Massion, 1994).

- *Controlled variables.* A body consists of many segments, each of which must be stabilized in relation to other segments, as well as to the external coordinate system. It is suggested that the central nervous system (CNS) subdivides this complex task into two simpler ones—maintenance of the body configuration and maintenance of the upright body orientation—and solves them separately (Horak and Macpherson, 1995). It is also suggested that the second task in terrestrial species can be solved by stabilizing the position of the center of mass (Massion, 1994, 1998).

- *Corrective motor responses.* Disturbances of the upright body posture may differ in their direction, amplitude, and so on. It is suggested that, to cope with these infinitely variable disturbances, a special strategy is used. This strategy includes a selection of the appropriate class of response (the muscle synergy) from a limited set of classes and regulation of the value of response (Horak and Nashner, 1986; Macpherson et al., 1997).

• *Capacity to stabilize different orientations* is a characteristic feature of the postural system in all species (Deliagina and Orlovsky, 1990; Fung and Macpherson, 1995; Horak and Macpherson, 1995; Orlovsky, 1991). It is suggested that gradual modifications of the stabilized orientation are due to a change of the control system set point, whereas the switches between strongly differing postures are caused by reconfiguration of the control network (Orlovsky, 1991).

• *Interaction of posture and movement.* Numerous studies in this field have shown that voluntary movements are accompanied (and often preceded) by specific postural adjustments that allow maintaining a balance (Massion, 1991, 1998; Massion and Dufossé, 1988). In locomoting animals and humans, postural modifications are incorporated into the basic locomotor pattern (Drew, 1991; Kably and Drew, 1998; Nashner and Forssberg, 1986; Orlovsky, 1972).

• *Recovery of postural control after trauma.* One of the most striking examples of the plasticity of motor control mechanisms is the recovery of postural control after the unilateral labyrinthectomy (UL) observed in various vertebrate species, from lamprey to man (Deliagina, 1997a; Dieringer, 1995; Smith and Curthoys, 1989). In all species, a complete initial loss of equilibrium is gradually compensated over time and finally disappears almost completely. It is suggested that recovery is due to several processes, including a substitution of vestibular input by visual and somatosensory inputs and recalibration of the system of sensory integration (Smith and Curthoys, 1989).

The conclusions and concepts listed previously relate mainly to the functional organization of postural control. Much less is known, however, about the organization and operation of the corresponding neuronal networks. This lag is caused primarily by methodological problems related to a specificity of the postural control system. A traditional experimental way to study complex neural mechanisms is to subdivide them into a number of smaller networks, each of which retains, at least partly, its normal function. This method is widely used to analyze, for example, the central pattern generators—that is, the neuronal networks capable of rhythmogenesis when isolated (Arshavsky et al., 1997; Marder and Calabrese, 1996; Selverston et al., 1997). However, this method is not applicable to the postural control system, which is a closed-loop system and therefore needs the integrity of the brainstem and spinal networks, as well as the presence of sensory feedback, for its normal function.

To overcome these difficulties, we turned to simpler "animal models," which present much better opportunities for the investigation of neuronal networks. Our hope was that such a basic problem as the nervous control of the antigravity behavior has similar solutions in the nervous systems of different species, and thus the results obtained on simpler models may have more general significance. We also developed several methods for studying postural activity in reduced preparations, as well as methods for artificially closing the sensory feedback loop (Deliagina et al., 1998a; Orlovsky et al., 1992; Zelenin et al., 2000). Two animal models were used in these studies: the marine mollusk *Clione limacina* and the lamprey, a lower vertebrate (cyclostome).

Studies on *Clione*

A relative simplicity of the central nervous system in *Clione,* together with a number of newly developed techniques, allowed us to obtain an almost complete description of the neuronal network responsible for postural control in this animal and to analyze its operation in detail under the conditions of open and closed feedback loop.

Postural Stabilization Is Based on Gravitational Reflexes

Clione is a planktonic animal. It is usually oriented vertically, with its head up (figure 4.2a). *Clione* swims upward or maintains itself at a particular depth by a continuous beating of two wings (Arshavsky et al., 1985, 1991; Satterlie et al., 1985). Any deviation from the vertical orientation evokes a corrective motor response, which includes tail bending and modification of wing beating (figure 4.2b) (Deliagina et al., 1998b; Panchin et al., 1995b). This response is aimed at restoration of the normal orientation.

The postural reflexes are driven by inputs from two statocysts. The statocyst in *Clione,* like that in other gastropod mollusks (Alkon, 1975; Janse, 1982), is a spheri-

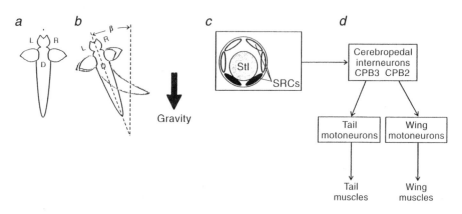

Figure 4.2   Postural reactions in *Clione* and principal elements of the postural network. *(a)* Normal orientation of *Clione. (b)* A deviation from this orientation to the left (β) evokes flexion of the tail to the right and asymmetry in wing beating. *(c)* Structure of the statocyst. The inner cavity of the statocyst is covered by the statocyst receptor cells (SRCs). The SRCs excited under the effect of the statolith (Stl) are shown in black. *(d)* The receptor cells send their axons to the cerebral ganglia and affect two groups of the cerebropedal interneurons: CPB2 and CPB3. These interneurons project to the pedal ganglia. The CPB3 interneurons affect the tail motoneurons, whereas the CPB2 interneurons affect the wing motoneurons. The tail motoneurons evoke bending of the tail, whereas the wing motoneurons have a double function: they evoke rhythmical locomotor oscillations of the wings and modulate these oscillations under the effect of gravitational input.

cal organ (figure 4.2c). About 10 statocyst receptor cells (SRCs) cover the inner surface of the statocyst wall (Tsirulis, 1974). A stonelike structure, the statolith (Stl) is located in the statocyst cavity. The SRCs are mechanoreceptors that are excited under the effect of the statolith when, because of a change in the organism's orientation, they are found in the lowermost position (Panchin et al., 1995b). After removal of both statocysts, *Clione* cannot maintain any definite orientation in space (Panchin et al., 1995b).

Postural Network

Paired recordings of cells in the statocyst and various ganglia of the CNS have revealed the main cellular groups constituting the postural network and their interconnections (Panchin et al., 1995a,b) (figure 4.2, c and d). The network is driven by SRCs, which exert excitatory and/or inhibitory action on the two groups of interneurons: CPB2 and CPB3. In their turn, the interneurons affect the tail and wing motoneurons responsible for elicitation of the corrective motor responses.

In the in vitro preparation consisting of the CNS and statocysts, we recorded responses to natural gravitational stimulation in different neurons that make up the postural network. The preparation was positioned in the chamber, the activity of motoneurons was recorded from their axons in the tail and wing nerves, and the activity of interneurons was recorded from their axons in the interganglionic commissure (figure 4.3, a and b) (Deliagina et al., 1998a, 1998b); the chamber was rotated in space in different planes, which corresponded to the sagittal sway, lateral sway, and horizontal roll (figure 4.3, c and d).

Spatial Zones of Activity of Different Neuron Groups

Figure 4.4 illustrates the responses of tail motoneurons and corresponding interneurons to rotation of the preparation throughout 360°, around its dorsoventral axis; the axis was situated horizontally (as in figure 4.3, C_2 and D_2). A deviation from the normal "head-up" orientation (0°) to the right (right sway) produced activation of motoneurons in the left tail nerve. Usually two to three (up to five) motoneurons in each tail nerve responded to gravitational stimulation. The size of their zone of activity varied from 135° to 210° in different experiments. The interneurons had zones of activity similar to the corresponding motoneurons (LSPC in figure 4.4).

Investigations of the responses of motoneurons in different tail nerves have shown that there are three major groups of motoneurons that differ greatly in their zones of activity. The top row of diagrams in figure 4.5 shows the angular zones for these three groups (T1, T2, and T3) revealed by rotation in different planes (i.e., by the sagittal sway, by the lateral sway, and by the horizontal roll). The zones are very wide (180°) and partially overlap.

The middle row of diagrams in figure 4.5 shows the angular zones of activity for different groups of wing motoneurons (W1-W4). They are slightly narrower than

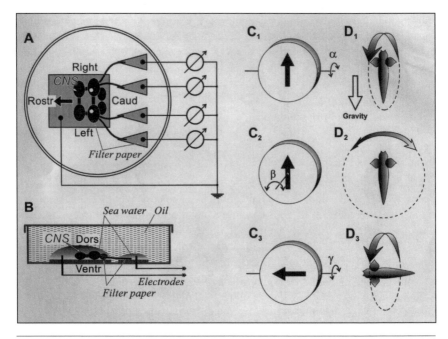

Figure 4.3   *(a, b)* Experimental arrangement for in vitro recording of the activity of motoneurons and interneurons. *(a)* The top view and *(b)* the side view of the recording chamber are shown. Five pieces of the filter paper soaked in sea water (electrodes) were positioned on the bottom of the chamber and isolated from each other by paraffin oil. The CNS isolated with the statocysts was positioned on the larger electrode with its dorsal side up. The arrow indicates the orientation of the rostral aspect of the CNS. The nerves (or the stumps of the transected subpedal commissure) were positioned on the smaller electrodes, which were connected with inputs of the amplifiers. The chamber was completely filled with oil and tightly closed. *(c, d)* Gravitational stimulation of statocysts. Three modes of rotation of the recording chamber are shown in C_1-C_3 (the arrows indicate the orientation of the rostral aspect of the CNS), whereas D_1-D_3 show the same three modes if they were applied to the whole animal. C_1 and D_1 = the sagittal sway characterized by the angle, α; C_2 and D_2 = the lateral sway characterized by the angle, β; C_3 and D_3 = the horizontal roll characterized by the angle, γ.

Reprinted from Deliagina, Orlovsky, Selverston, et al. 1999.

the zones of tail motoneurons (100° against 180°). Another difference is that the wing motoneurons could not be activated by the sagittal sway.

The bottom row of diagrams in figure 4.5 shows the zones of activity of four groups of interneurons (IN1-IN4) controlling the tail motoneurons. There is a striking similarity between the zones of IN1-IN3 groups of interneurons and the zones of T1-T3 groups of motoneurons, both in zone position and zone width. This similarity suggests that there is no essential transformation of the gravitational information when it passes from inter- to motoneurons and that formation of the zones

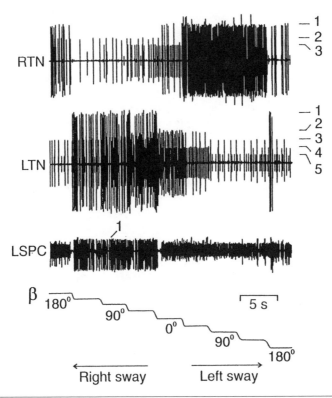

Figure 4.4   Responses of motoneurons and interneurons to rotation of the preparation in the frontal plane (as in figure 4.3, C_2 and D_2). Motoneurons in the right tail nerve (RTN, units 1-3) and in the left tail nerve (LTN, units 1-5) were activated preferentially with the contralateral sway. An interneuron with larger spikes, recorded in the left stump of the subpedal commissure (LSPC, unit 1), was activated with the right sway. The angle of sway (β) was changed in 45° steps.

Reprinted from Deliagina, Arshavsky, and Orlovsky 1998.

occurs at the preceding stage (i.e., when information passes from SRCs to interneurons). The group IN4, however, has no counterpart among the motoneurons; its zone coincides with the area of silence of the T1 group, suggesting that IN4 interneurons exert an inhibitory action on T1 motoneurons.

The one-dimensional zones for the T1-T3 and W1-W4 groups of motoneurons (figure 4.5) presented a basis for their three-dimensional reconstruction, shown in figure 4.6, a-d. On these graphs, the orientation of *Clione* in relation to the gravitational force is represented by the radius-vector originating from the center of the sphere (shown in figure 4.6a). The angles α and β show deviation of *Clione* from the vertical orientation for the sagittal and lateral sway, respectively. The arcs drawn on the sphere by thick lines show the width and position of the one-dimensional zones revealed by rotation in different planes; the thin lines connect the extreme points of

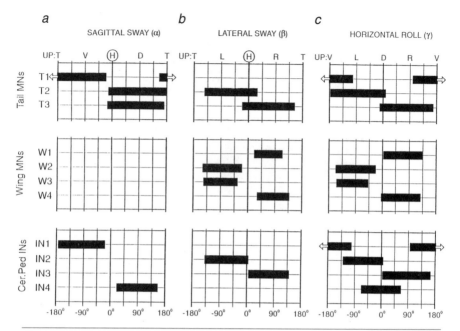

Figure 4.5   Angular zones of activity for different groups of neurons averaged over all experiments. The neurons were tested by the *(a)* sagittal sway, *(b)* lateral sway, and *(c)* horizontal roll. The horizontal bars indicate the average width of the zones. Designations for different positions of the preparation are given in terms of the position of the whole animal: V = ventral side up, D = dorsal side up, H = head up, T = tail up, R = right side up, L = left side up, MNs = motoneurons, INs = interneurons, Cer. Ped = cerebropedal. The normally stabilized orientation (H) is marked by a circle.

the one-dimensional zones to show the presumed borders of the three-dimensional zones. One can see that the T1-T3 zones of the tail motoneurons are very wide, each of them occupying approximately half of the sphere, and the zones of different groups overlap considerably with each other. In contrast, the W1-W4 zones of wing motoneurons are narrower and do not overlap.

Correlation Between Motor Responses Evoked by Different Groups of Motoneurons and Their Zones of Activity

The motor effects produced by different groups of tail and wing motoneurons can be estimated on the basis of experiments with stimulation or transection of the corresponding nerves (Panchin et al., 1995a). These effects are illustrated in figure 4.6, e and f. The T2 and T3 groups of tail motoneurons elicit a tail flexion to the left and to the right, respectively, with some dorsal component. The T1 group elicits a ventral tail flexion. The W1 and W2 groups of wing motoneurons increase the amplitude of locomotor beating of the left and right wings, respectively. Finally, the

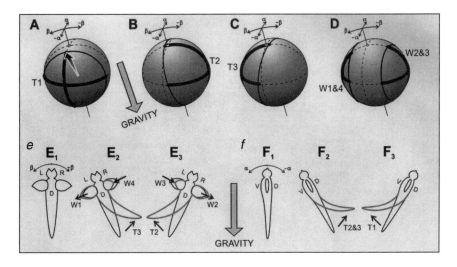

Figure 4.6 Spatial zones of activity of different groups of tail and wing motoneurons and corrective motor responses evoked by these groups. *(a-d)* Zones of activity of T1-T3 groups of tail motoneurons and W1-W4 groups of wing motoneurons. The radius-vector in *(a)* represents orientation of *Clione* in the gravity field. The arc on the sphere drawn by thick lines shows the angular width and position of the 1-dimensional zones (based on data presented in figure 4.5). *(e, f)* E_1-E_3 and F_1-F_3: motor responses elicited by activation of different groups of motoneurons when *Clione* is deviated from the normal position (the normal orientation is shown in E_1 and F_1). The effect of activation of particular neuron groups on the configuration of the tail and wings is shown by gray lines and marked by arrows.

Reprinted from Deliagina, Orlovsky, Selverston, et al. 1999.

W3 and W4 groups cause a wing retraction and thus decrease the amplitude of beating of the left and right wings, respectively.

Comparison of the zones of activity of different motoneuron groups (figure 4.6, a-d) and the motor effects they produce (figure 4.6, e and f) shows that at any spatial orientation of *Clione*, the gravitational reflexes are aimed at restoration of the normal orientation. For example, the left sway (E_2) elicits tail flexion to the right. The deviated tail, like the rudder of a boat, will cause rotation of *Clione* toward the vertical. Similarly, an increase of oscillations in the left wing and a decrease in the right wing will produce a torque rotating *Clione* toward the vertical. The wing reactions supplement the main corrective response (i.e., the tail flexion) only in a part of space where the lateral component of tilt is strongly expressed (figure 4.6d).

Activity of Postural Network During Stabilization of Spatial Orientation

To investigate the activity of the postural network under closed-loop conditions, we used a new method that combined in vitro and robotics approaches (figure

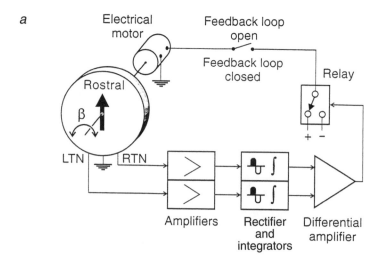

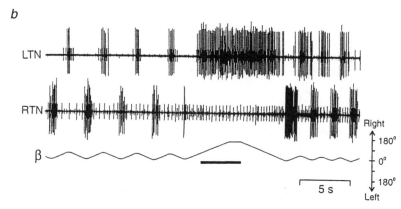

Figure 4.7 *(a)* Experimental arrangement for automatic stabilization of the orientation of the CNS in space. Signals (action potentials) from the left and right tail nerves (figure 4.3, a and b) were processed by the electronic circuit and then used to drive an electrical motor that rotated the chamber with the preparation. *(b)* Activity of motoneurons from the left and right tail nerves (LTN and RTN) recorded under closed feedback loop conditions. The postural network stabilized the "head-up" orientation, with small oscillations around it. A horizontal bar indicates the period when the feedback loop was opened, and a large deviation from the stabilized orientation was evoked by a continuously rotating motor. After the disturbance had terminated and the loop closed again, the system restored a normal orientation.

Reprinted from Deliagina, Arshavsky, and Orlovsky 1998.

4.7) (Deliagina et al., 1998a). The CNS-statocysts preparation was positioned in the chamber (as in figure 4.3, a and b). We used the output signals of the network (i.e., the electrical discharges of tail motoneurons) to control an electrical motor rotating the preparation. The stabilization of orientation in the frontal plane (as in figure 4.3, C_2 and D_2) was investigated first. The output signals were taken from the left and right tail nerves (LTN and RTN), which contain the axons of the T2 and T3 groups of tail motoneurons. By means of an electrical circuit (figure 4.7a), these signals were amplified, rectified, and integrated. They were then passed to inputs of the differential amplifier, which, through a relay, produced a reverse in the rotation of the electrical motor. If the spike frequency in the left nerve prevailed over that in the right nerve, rotation was to the left; in the opposite case, it was to the right.

Figure 4.7b shows the effect of closing the feedback loop. The network stabilized its own spatial orientation, which corresponded to the head-up orientation of *Clione*, with small oscillations around it. The oscillations were caused by excursions of the system between the zones of activity of the T2 and T3 groups of motoneurons. The system was resistant to externally applied disturbances of the orientation of the preparation in space. As shown in figure 4.7b, a large (~170°) imposed deviation from the normal orientation, caused by rotation of the preparation when the feedback loop was open, was rapidly compensated for after the loop was closed again. A similar result was obtained when the activity of the T1 and T2 groups of motoneurons was used as the driving signals to stabilize the orientation in the sagittal plane (as in figure 4.3, C_1 and D_1). These experiments confirmed the theory (Deliagina et al., 1998b) that postural corrective responses are caused by three groups of motoneurons that elicit tail flexion in three different directions, and these groups are sufficient for restoration of the normal position after any postural disturbances.

Reconfiguration of Postural Network Underlying Stabilization of Different Postures

Clione has been shown to stabilize its head-up orientation only at lower temperatures. At higher temperatures (15-20°C), it stabilizes the head-down orientation and swims downward, away from the warmer water layers (Panchin et al., 1995a). We found that postural activity of the in vitro preparations also depended on temperature (figure 4.8). With the closed feedback loop, the head-up orientation was stabilized at 10°C (figure 4.8a). After the temperature was raised to 20°C, the system stabilized the head-down orientation (figure 4.8b). A basis for these modifications was the temperature-dependent reconfiguration of the network revealed by recording the gravitational responses in inter- and motoneurons under open-loop conditions. We found that at 10°C, the SRCs, excited by gravitational input, activated the contralateral reflex chain (figure 4.8, c and e). At 20°C, however, the same SRCs activated the ipsilateral chain (figure 4.8, d and f). Thus, the postural network is not a hardwired circuit; it can be modified to such an extent that a nearly "new" network is formed, with a reversed response to the same gravitational input. Modifications of networks by different modulatory inputs have been described for a number of motor behaviors (Harris-Warrick and Marder, 1991; Selverston, 1993, 1995), including postural control (Arshavsky et al., 1993).

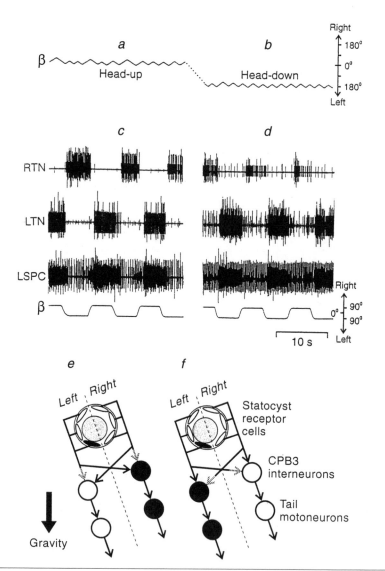

Figure 4.8 Activity of postural network depends on temperature. *(a, b)* Reversal of the stabilized orientation (closed feedback loop). At 10°C, *(a)* the system stabilized the head-up orientation. With a temperature rise up to 20°C, *(b)* the system switched to stabilization of the head-down orientation (this was preceded by a period of instability indicated by the dotted line). *(c, d)* Reversal of gravitational reflexes (opened feedback loop). At 10°C, *(c)* the motoneurons (in RTN and LTN) and the interneurons (in the stump of the left subpedal commissure) were activated with the contralateral sway. At 20°C, *(d)* they were activated with the ipsilateral sway. *(e, f)* Temperature-dependent reconfiguration of the postural network. Neurons activated with the left sway are shown in black. At lower temperatures, the SRC, excited by gravitational input, activates the contralateral reflex chain. At higher temperatures, the same SRC activates the ipsilateral chain.

Reprinted from Deliagina, Arshavsky, and Orlovsky 1998.

Studies on Lamprey

The lamprey (a lower vertebrate, cyclostome) has the basic central nervous system structure similar to that in higher vertebrates (Kappers, 1936) but presents many more opportunities for analytical study of motor control mechanisms and of the postural control system in particular (Grillner et al., 1995; Macpherson et al., 1997). This is mainly because the brainstem–spinal cord preparation remains in a functional state for many hours after its isolation and exhibits neuronal correlates of the spatial orientation behavior (Orlovsky et al., 1992).

Postural Stabilization Is Based on Vestibular Reflexes

When swimming, the lamprey is normally oriented with its dorsal side up (figure 4.9, a and b). Any deviation from this orientation (roll tilt, α; figure 4.9c) evokes a corrective motor response aimed at restoration of the normal orientation. This response may include a lateral flexion of the ventrally deviated tail, a lateral deviation of the dorsal fin, and a body twisting (Ullén et al., 1995a). These movements all generate a torque rotating the lamprey around its longitudinal axis in a direction opposite to the initial tilt.

Postural corrective reflexes in the lamprey are driven by vestibular input. After a unilateral or bilateral labyrinthectomy, the lamprey is not able to maintain a particular orientation in space (de Burlet and Versteegh, 1930; Deliagina, 1997a; Ullén et al.,

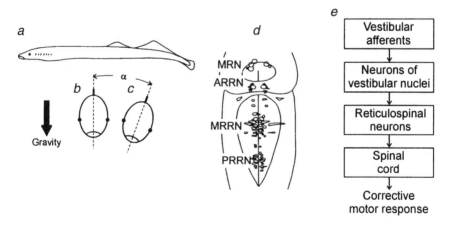

Figure 4.9 Postural orientation in the lamprey and principal elements of the postural network. *(a, b)* Normal orientation of the lamprey (side and front views). *(c)* A deviation from this orientation (roll tilt, α) evokes a set of corrective motor responses aimed at restoration of the normal orientation. *(d)* Four reticular nuclei of the brainstem giving rise to reticulospinal pathways: MRN = the mesencephalic reticular nucleus; ARRN, MRRN, and PRRN = the anterior, middle, and posterior rhombencephalic reticular nuclei. *(e)* Principal neuron groups involved in postural control in the lamprey.

1995a). In contrast to vestibular input, visual input exerts only a modulatory effect on the postural orientation—it elicits a lateral tilt toward the more illuminated eye (dorsal light response; see inset in figure 4.11c) (Ullén et al., 1995b; von Holst, 1935).

Brainstem Mechanisms for Postural Control

As in other vertebrates, the basic neural mechanisms for postural control in the lamprey are located in the brainstem and in the spinal cord. The brainstem mechanisms process and integrate vestibular and visual signals and send commands to the spinal cord. In response to these commands, the spinal mechanisms generate corrective motor responses. Commands for postural corrections can be transmitted via the reticulospinal (RS), vestibulospinal, and propriospinal pathways (Rouse and McClellan, 1997; Rovainen, 1979). Deliagina et al. (1993) suggested that the RS pathways are most important for transmitting postural commands. The two bilaterally symmetrical RS pathways originate from the neurons of four reticular nuclei of the brainstem (figure 4.9d) and reach even the most caudal spinal segments (Nieuwenhuys, 1972; Rovainen, 1979). They exert diverse effects on the ipsilateral and contralateral inter- and motoneurons (Ohta and Grillner, 1989; Zelenin et al., 2001).

The RS neurons receive input from vestibular afferents via the interneurons located in the vestibular nuclei of the lateral medulla (figure 4.9e). We studied the input and output signals in this network by recording responses to natural vestibular stimulation in two classes of neurons: the vestibular afferents and the RS neurons (Deliagina et al., 1992a, 1992b). The in vitro preparation for recording these responses, shown in figure 4.10a (Orlovsky et al., 1992), consists of the brainstem isolated with the vestibular organs and eyes. The preparation could be rotated in space in the frontal or sagittal plane; here the responses to roll tilt (α) will be considered. Vestibular input could be combined with visual input (eye illumination). These experiments have shown that vestibular afferents are activated with the ipsilateral roll tilt (figure 4.10b). Since these afferents cause excitation of the RS neurons on the contralateral side, the RS neurons become activated with the contralateral roll tilt (figure 4.10c). They also receive visual input and become activated with illumination of the ipsilateral eye (figure 4.10d).

Conceptual Model of the Roll Control System

These experimental findings led to formulation of the conceptual model of the roll control system in the lamprey (figure 4.11a) (Deliagina, 1997a, 1997b; Deliagina et al., 1993). The left and right groups of RS neurons, RS(L) and RS(R), receive excitatory input from the otolith organ of the contralateral labyrinth; they also receive excitatory input from the ipsilateral eye. The RS neurons exert an action on spinal motoneurons (Ohta and Grillner, 1989; Zelenin et al., 2001), which presents a basis for the generation of postural corrections (i.e., rotation of the animal in the direction opposite to the initial roll tilt) (Ullén et al., 1995a).

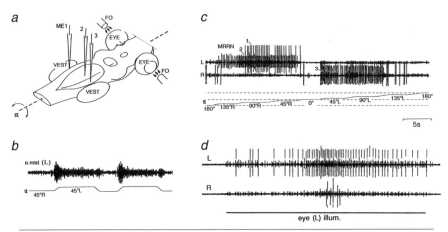

Figure 4.10 *(a)* Experimental arrangement for in vitro investigation of the postural network in the lamprey. The brainstem was isolated together with the vestibular organs and eyes. For vestibular stimulation, the preparation was rotated around its longitudinal axis (roll tilt, α); for visual stimulation, one of the eyes was illuminated by the fiber-optic (FO) system. Responses in brainstem neurons were recorded extracellularly by microelectrodes (ME1-3). *(b)* A mass activity in vestibular afferents was evoked by the ipsilateral roll tilt. *(c, d)* Activity of RS neurons was recorded from the left (L) and right (R) reticular nuclei. The neurons were activated by the *(c)* contralateral roll tilt and *(d)* ipsilateral eye illumination.

Figure 4.10c reprinted from Deliagina, Orlovsky, Grillner, et al. 1992. Figure 4.10d reprinted from Deliagina, Orlovsky, Grillner, et al. 1993.

Figure 4.11b shows schematically the roll-dependent activity in the two antagonistic groups of RS neurons and the corrective motor response they elicit (arrows). The effects of the two groups are equal to each other at 0° (the dorsal-side-up orientation), and no corrective motor response will occur at this position. Any deviation from the equilibrium point will elicit a corrective motor response aimed at restoration of the normal orientation.

The model shown in figure 4.11a, despite its simplicity, can account for several phenomena related to postural control in the lamprey. One of these is the dorsal light response (i.e., the roll tilt toward the illuminated eye). In the model, visual input excites the ipsilateral group of RS neurons, which will cause a bias in their response to roll tilt (figure 4.11c). As a result, intersection of the RS(R) and RS(L) curves (i.e., the equilibrium point of the system) will be shifted toward the illuminated side, and this new position will be stabilized by the gravitational orientation system (inset in figure 4.11c).

Another phenomenon that can be explained by the model is the postural disorders caused by the unilateral labyrinthectomy (UL). Deprived of one labyrinth (figure 4.12a), the swimming lamprey rotates continuously around its longitudinal axis toward the lesioned side (inset in figure 4.12c) (Deliagina 1997a; Ullén et al., 1995b). According to the model (figure 4.12a), the group of RS neurons located on the lesioned side is excited at any roll angle by input from the intact, contralateral

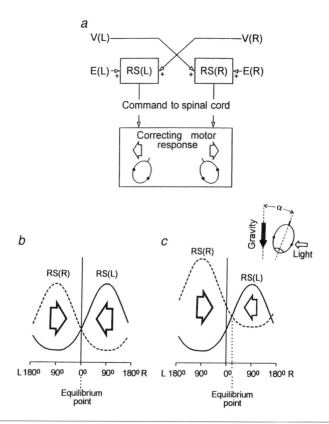

Figure 4.11   *(a)* A conceptual model of the postural control system in the lamprey (see text for explanations). *(b)* Operation of the system when driven only by vestibular inputs. The curves represent responses in two subpopulations of reticulospinal neurons, RS(L) and RS(R), to left (L) and to right (R) roll tilt (abscissa is roll tilt angle; ordinate is activity of RS neurons). The RS(L) and RS(R) are activated by vestibular input with contralateral roll tilt. Correcting motor responses, evoked by the predominating subpopulation of RS neurons, are indicated by arrows. The system has an equilibrium point at 0° (the dorsal-side-up orientation). *(c)* Illumination of the right eye causes an upward transition of the RS(R) curve and a shift of the equilibrium point. This results in inclination of the animal toward the light (dorsal light response; see inset).

labyrinth, while the other group is silent or exhibits low activity (figure 4.12c). The RS(R) and RS(L) curves do not intersect, the system has no equilibrium point, and the prevailing RS group causes continuous rolling.

Vestibular Compensation

The UL causes severe postural disorders in all classes of vertebrates. Recovery after UL is a slow process (Smith and Curthoys, 1989), lasting for several weeks in the

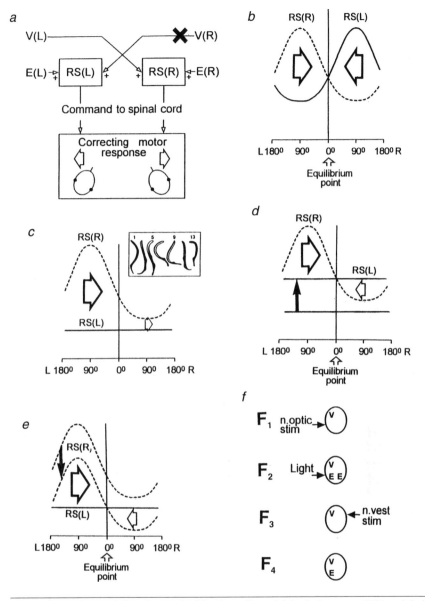

Figure 4.12 Postural disorders evoked by the unilateral labyrinthectomy (UL) and their compensation. *(a)* A model of the postural control system with eliminated input from the right labyrinth. *(b, c)* Operation of the system *(b)* before UL and *(c)* after UL. After UL, the RS(L) and RS(R) curves do not intersect, and the system has no equilibrium point; the lamprey rotates continuously (inset). *(d)* The equilibrium can be restored by means of tonic excitatory drive to RS(L). *(f)* This can be achieved by n. optic stimulation (F_1), by eye illumination (F_2), or by n. vestibular stimulation (F_3). *(e)* The equilibrium can be restored by reduction of tonic excitatory drive to RS(R) due to, for example, removal of an eye (F_4).

lamprey (Deliagina, 1997a). The model (figure 4.12a) provides a key to understanding both the origin of postural disorders and the recovery of postural control (vestibular compensation). It is evident that for the two curves (figure 4.12c) to be intersected, either the lower curve must be translated upward (figure 4.12d) or the upper curve translated downward (figure 4.12e). The upward translation of the lower curve can be performed experimentally in different ways, all of which are based on supplying the RS neurons that lost their vestibular input with a tonic excitatory drive. These ways include electrical stimulation of the optic nerve on the side contralateral to the UL, illumination of the eye on that side, and stimulation of the vestibular nerve on the UL side (figure 4.12f, F_1-F_3). The downward translation of the upper curve can be produced by depriving the "stronger" group of RS neurons of visual input, which can be done, for example, by removal of the corresponding eye (figure 4.12, F_4). All these methods were tested and found very efficient (Deliagina, 1997a, 1997b). For example, stimulation of the optic nerve (via an implanted electrode) led to immediate termination of rolling, and the animal swam normally as long as the stimulation continued (Deliagina, 1997b).

One of the methods for compensation of the vestibular deficit, developed for the lamprey (i.e., electrical stimulation of the vestibular nerve on the UL side), was subsequently used in the rat (Deliagina et al., 1997). Normally, UL in the rat elicits several different symptoms (motor disorders) that gradually diminish with time. They include head roll tilt, body twisting, extension of the contralateral limbs, and rolling. Stimulation of the vestibular nerve instantaneously abolished these symptoms, which strongly suggests a similarity in the functional organization of postural mechanisms in the evolutionary remote species.

RS Activity in Intact Lampreys

Our knowledge of the function of the RS system in postural control was obtained in in vitro experiments. However, the in vitro preparation differs from the intact CNS primarily in that a number of inputs to the brainstem postural network are lacking (e.g., inputs from the higher brain centers, inputs from the spinal cord, and inputs from the cranial nerves). This can modify, to some extent, the processing of vestibular and visual information in the network. To elucidate the activity of the RS system under normal conditions, we recorded the activity of RS neurons in intact lampreys by means of implanted electrodes (Deliagina et al., 2000; Deliagina and Fagerstedt, 2000). Two arrays of electrodes (two or four electrodes each) were positioned on the dorsal surface of the spinal cord at different rostrocaudal levels; one of them is shown in figure 4.13a. By comparing the spike amplitude and the time of the spike occurrence on different electrodes, we could separate discharges in individual axons (figure 4.14a), estimate the axonal position in the spinal cord, and measure its conduction velocity.

Figure 4.13b shows the setup for studying vestibular and visual responses in the RS neurons of the intact lamprey. Spontaneous movements of the animal were restrained by positioning it within a tube; a roll tilt (α) could be produced by rotating

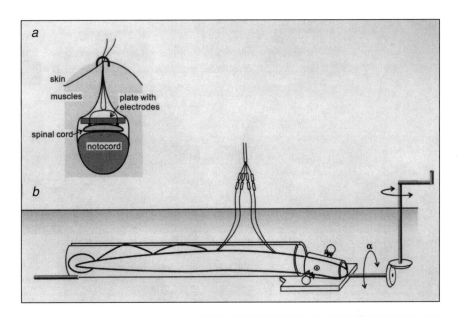

Figure 4.13   Experimental arrangement for investigation of RS neurons in intact lampreys. *(a)* Activity was recorded by implanted electrodes. Two plates, one plate with two electrodes (as in *[a]*) and one with four electrodes, were implanted at two different rostrocaudal levels and positioned on the dorsal aspect of the spinal cord. The electrodes recorded activity in the larger RS axons. The position of each individual axon in relation to the electrodes could be estimated by comparing the spike amplitude on the electrodes. *(b)* Setup for rotating the lamprey. The animal was positioned in the tube fastened to the rotating platform. The animal's eyes could be illuminated by the fiber-optic systems. The roll tilt (α) could be caused by rotating the platform.

Reprinted from Deliagina and Fagerstedt 2000.

the tube. The implanted electrodes recorded the mass activity in the RS pathways. Here we focus on the activity of larger RS neurons—the Mauthner cells, Müller cells, and V cells. Activity of individual axons was extracted from the mass activity by means of the spike-sorting program (figure 4.14a).

The RS neurons in the intact lamprey were found to be activated by the contralateral tilt and by the ipsilateral eye illumination (figure 4.14a), as demonstrated earlier in the in vitro experiments. Responses in intact animals differed, however, in that they had a stronger dynamic component and their static component was more sensitive to the direction of rotation—it was much stronger with rotation in one direction than in the other (figure 4.14c). Two groups of RS neurons could be distinguished. Static responses were more pronounced in the group 1 neurons with more lateral axons (most likely, V cells) (figure 4.14a), whereas dynamic responses were more pronounced in the group 2 neurons with medial axons (B and M cells).

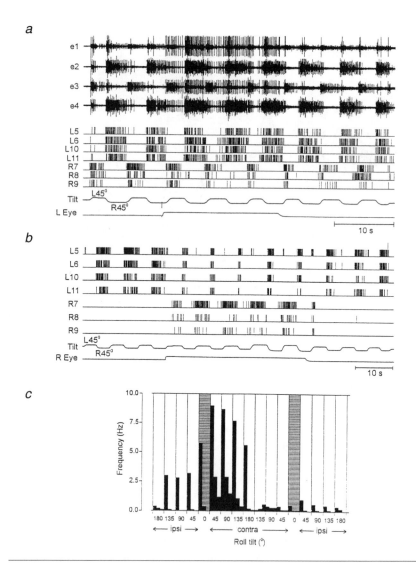

Figure 4.14 Responses of RS neurons in the intact lamprey to vestibular and visual inputs. (a) The four upper traces show the mass activity recorded by the four-electrode array (e1-e4). By means of the spike-sorting program (Datapack 3, Run Technologies Inc.), the activity of individual RS axons was extracted from the mass activity. Shown here is the activity in the group 1 axons (L5, L6, L10, L11 on the left side and R7, R8, R9 on the right side), which exhibited both dynamic and static responses to the contralateral roll tilt. (b) The same neurons as in (a) were tested after removal of the left labyrinth. In R7-R9, the vestibular responses disappeared; they were restored when the right eye was illuminated. (c) Responses of group 1 neurons to rotation through 360° performed by successive 45° steps (as in figure 4.4), clockwise (first turn) and counterclockwise (second turn). Responses were averaged over 42 neurons recorded in six experiments. ·

Modified from Deliagina and Fagerstedt (2000).

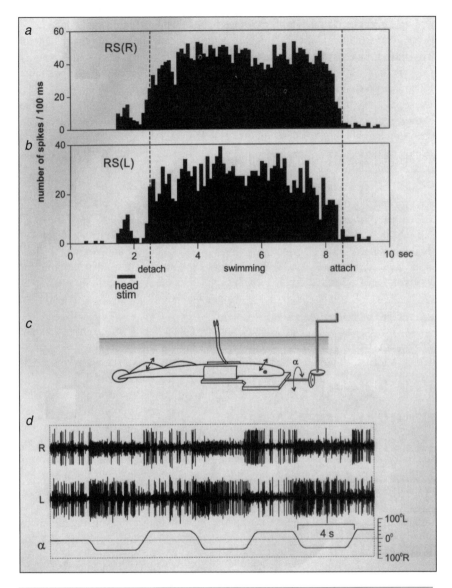

Figure 4.15 Participation of RS system in the control of locomotion and posture. *(a, b)* Mass activity in the right *(a)* and left *(b)* RS pathways recorded by implanted electrodes in the freely behaving lamprey. Initially, the animal was in a quiescent state, attached to the substratum by its sucker mouth. Swimming was then evoked by tactile stimulation. *(c, d)* Interaction of postural and locomotor commands. *(c)* A setup for studying postural responses during swimming. The animal was restrained so that it could perform locomotor movements but could not itself correct its postural orientation. *(d)* Vestibular responses in the right (R) and left (L) RS axons during locomotion.

Reprinted from Deliagina, Zelenin, Fagerstedt, et al. 2000.

Experiments on the whole animal have shown that, in addition to the main input from the contralateral labyrinth, the RS neurons receive input from the ipsilateral labyrinth. Figure 4.14a shows responses to the left/right 45° tilts in the group 1 neurons of the intact lamprey. After removal of the left labyrinth, responses in the group 1 neurons located on the right side disappeared (left part of figure 4.14b). These responses could be restored, however, by illuminating the right eye (middle part of figure 4.14b). The ipsilateral vestibular input to RS neurons is weaker than the contralateral input. One can suggest, however, that the response to this input is potentiated during recovery after UL until this input is able to substitute the main contralateral input.

Interaction of Postural and Locomotor Systems

The RS command system in the lamprey is a multifunctional system responsible for the control of swimming, for postural corrections, for steering, and the like (Grillner et al., 1995; Wannier et al., 1998). Figure 4.15, a and b, shows that initiation of swimming in a quiescent, freely behaving lamprey, caused by tactile stimulation of the head, was accompanied by an enormous increase of the bilateral RS activity (Deliagina et al., 2000).

To study interactions of the postural and locomotor commands conveyed by the RS system, we used an intact lamprey whose movements were partly restrained (figure 4.15c); the anterior and posterior parts of the body could perform locomotor oscillations, but the lamprey could not perform postural corrections because of fixation of the midbody part. Figure 4.15d shows the RS activity recorded during swimming by two implanted electrodes (L and R). A clear-cut rhythmical modulation of the activity of RS neurons is evident. This modulation was linked to the locomotor cycle and was caused by the efference copy signals coming from the spinal locomotor network (Kasicki et al., 1989). Lateral tilts of the animal evoked vestibular responses in the RS neurons; these responses were superimposed on the swim-rhythm-related modulation of the neuron discharge. Thus, the commands for postural corrections are subjected to rhythmical gaiting in the brainstem and therefore can affect the spinal mechanisms in one phase of the swim cycle but not in the other phase. This may explain the observations that postural corrections do not disturb the ongoing locomotor pattern but rather are accurately incorporated into the locomotor rhythm (Hirschfeld and Forssberg, 1991; Nashner and Forssberg, 1986; Orlovsky, 1972).

Activity of Postural Network With Artificially Closed Feedback Loop

To study postural network activity under near-normal conditions, the method of artificial feedback initially used with the *Clione* model (figure 4.7a) was further elaborated (Zelenin et al., 2000). The intact lamprey was mounted on the platform (as in figure 4.13b), and its movements were restrained. The platform was rotated

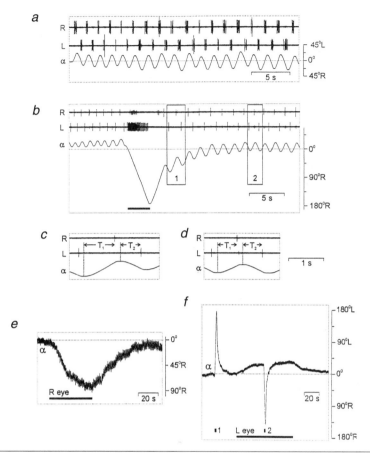

Figure 4.16 Automatic stabilization of postural orientation in the lamprey with artificially closed feedback loop. Activity in the left and right RS pathways (L and R in *(a)* through *(d)*) was recorded by means of implanted electrodes, processed by an electronic circuit (as in figure 4.7a), and then used to control an electrical motor rotating the lamprey. *(a, b)* The system stabilized the dorsal-side-up orientation, with small periodical oscillations around it. *(a)* Stabilization of the normal orientation when the amplification in the control circuit was low; *(b)* the same with high amplification in the control circuit. The system rapidly compensated for postural disturbances. The horizontal bar indicates a period when the loop was opened and a postural disturbance was evoked by the motor rotation. *(c, d)* Parts of recording indicated by rectangles in *(b)* (1 and 2) are shown with higher time resolution. Postural stabilization was based on the dynamic vestibular reactions in RS neurons caused by oscillations. Temporal characteristics of these reactions, including their latency, were position-dependent. *(e, f)* Simulation of the dorsal light response in the lamprey with artificially closed feedback loop. *(e)* Response to illumination of the right eye. *(f)* Postural corrections under two conditions: with no eye illumination and on the background of the dorsal light response caused by the left eye illumination. Bars 1 and 2 indicate the periods when the loop was opened and postural disturbances (~180° in amplitude) were caused by the experimenter.

Reprinted from Zelenin, Deliagina, Grillner, et al. 2000.

by an electrical motor to produce a roll tilt of the animal. Discharges of the larger RS axons were recorded separately from the left and right sides of the spinal cord; the two sides were electrically isolated from each other by a longitudinal wall attached to the plate with the electrodes. The signals from the electrodes were processed by an electronic circuit (as in figure 4.7a) and used to control the motor. The direction of rotation was determined by the prevailing (left or right) signal. Thus, the spinal networks, together with locomotor organs, were replaced by a robotic system.

Under open-loop conditions, the majority of RS neurons were activated with the contralateral roll tilt (as in figure 4.14a). Under closed-loop conditions, the system stabilized a nearly normal orientation of the lamprey, with periodic oscillations around this value (figure 4.16a). The oscillations were caused by excursions of the system between the angular zones of activation of the antagonistic groups of RS neurons. The bursts of activity of the left and right RS neurons caused a reversal of motor rotation. With a higher gain in the electronic circuit, a reversal could be caused by a single (first) spike generated by the neuron most sensitive to roll tilt (figure 4.16b).

Any disturbance of a postural orientation was compensated for rapidly by a system (figure 4.16b). The return to the normal orientation was not smooth but oscillatory, however, due to the alternating dynamic vestibular responses in the left and right RS neurons resulting in a reversal of rotation. This striking result can be explained by the observation that the responses in RS neurons occurred with different delay in relation to the movement reversal. As shown in figure 4.16c, the delay for the "correct" response (i.e., for activation of the left RS neuron [T2]) was shorter than for the "incorrect" response (i.e., for activation of the right RS neuron [T1]). In contrast, during oscillations around 0°, the delays T1 and T2 were equal to each other (figure 4.16d). The postural stabilization was thus based on the dynamic vestibular reactions in RS neurons; the characteristics of these reactions were, however, position-dependent.

Modification of Body Orientation Caused by Visual Input

Under open-loop conditions, illumination of one eye evoked activation of ipsilateral RS neurons (similar to that shown in figure 4.10d). Under closed-loop conditions, illumination of one eye evoked a roll tilt toward the illuminated eye (figure 4.16e) (i.e., the behavior [dorsal light response] observed in freely behaving animals (Ullén et al., 1995b).

A new position resulting from the eye illumination was actively stabilized by the system, as demonstrated by perturbing the postural equilibrium. As shown in figure 4.16f, an externally imposed roll tilt (bar 1) was rapidly compensated, and the system returned to the initial orientation (0°). A consistent roll tilt of 35° was then evoked by eye illumination. On this background, the externally imposed tilt (bar 2) was also rapidly compensated. In this case, however, the system returned not to 0° but to the currently stabilized orientation (i.e., 35° tilt to the left).

One can thus conclude that visual input causes a shift of the equilibrium point in the vestibular-driven postural control system. A transition from one stabilized orientation to another (figure 4.16e) can thus be considered an active movement caused by a displacement of the equilibrium point in the control system (Feldman, 1986).

Conclusion

In the introductory paragraphs, a few principal points concerning functional organization of the postural control system were considered. Studies on *Clione* and lamprey allowed us to better understand the functional organization of the system, as well as to reveal some of the corresponding neuronal mechanisms.

1. *Stabilization of body orientation in any plane is based on antagonistic postural reflexes.* Both in the lamprey and *Clione*, the postural system stabilizes the body orientation at which the antagonistic vestibular reflexes are equal to each other (figure 4.11). A subtraction of the two reflexes occurs at a low level—in the spinal cord (lamprey) or even in the motor system (*Clione*). These findings suggest that, in these animals, there are no signals anywhere in the postural control system that are directly monitoring the animal's orientation in space.

2. *Processing and integration of sensory inputs.* Experiments on *Clione* have shown that the principal stage in the processing of gravitational signals is the formation of spatial zones of gravitational reflexes. This formation is based on the convergence of inputs from specific gravitational receptors on the interneurons mediating the reflexes.

A neuronal substrate for integration of sensory inputs of different modalities was revealed in the lamprey. The integration was found to be based on the convergence of two inputs—vestibular and visual—on the RS neurons (figure 4.10, c and d).

3. *Controlled variables.* Both in the lamprey and in *Clione*, vestibular input allows stabilization of the head orientation in the gravitational field. Stabilization of the whole-body orientation by the vestibular-driven mechanisms, however, requires the activity of special nervous mechanisms that maintain a proper head-trunk configuration. In the lamprey, the rectilinear head-trunk configuration is presumably maintained due to the segmental reflexes driven by the intraspinal stretch receptors (Viana di Prisco et al., 1990).

4. *Corrective motor responses.* Postural corrections in *Clione* are caused by three basic motor patterns that combine the tail and wing movements (gross synergies). Each of these synergies occurs within a specific zone of deviation of *Clione* from the vertical. The zones are very wide and partially overlap (figure 4.6).

5. *Stabilization of different orientations.* Experiments on the lamprey presented direct evidence that a gradual change of postural orientation is caused by a displacement of the equilibrium point in the control system under the effect of tonic drive to RS neurons (figure 4.11c). Experiments on *Clione* have demonstrated that switching

between the two strongly differing postural orientations is based on a reconfiguration of the postural network (figure 4.8).

6. *Interaction of posture and movement.* In both *Clione* and lamprey, postural corrections are incorporated into the ongoing locomotor rhythm. In *Clione*, this is achieved by convergence of the postural commands and the commands from the locomotor networks on the same wing motoneurons (Deliagina et al., 1998b). This allows the postural corrections to occur without a disturbance of the basic locomotor rhythm. In the lamprey, the same effect is achieved due to the gating mechanism in the RS neurons (figure 4.15d). The mechanism is driven by the efference copy signals from the spinal locomotor networks; it allows a transmission of postural corrections only in a certain phase of the locomotor cycle.

7. *Recovery of postural control after UL.* Of the two suggested mechanisms for recovery after UL (vestibular compensation), that is, a recalibration of the existing pathways and a substitution of vestibular input by visual input (Smith and Curthoys, 1989), the first mechanism seems more likely. We found that in the RS neurons, deprived of their main vestibular input from the contralateral labyrinth, the vestibular responses could be restored in the presence of tonic excitatory visual input (figure 4.14b). Thus, one can suggest that those plastic changes in the postural network that increase the excitability of RS neurons will lead to the recovery of postural control. The other presumed mechanism for compensation (i.e., substitution of damaged vestibular input by visual input) is unlikely since visuopostural reflexes in the lamprey are not fast enough to be responsible for postural stabilization (figure 4.16e).

8. *Redundancy in supraspinal commands for postural corrections.* Experiments on the lamprey with artificial feedback have shown that a small portion of the signals transmitted from the brainstem to the spinal cord is sufficient for control of body orientation in the roll plane. In some cases, the signals transmitted by one single right and one single left RS axon were sufficient (figure 4.16b). These signals contain information necessary for stabilization of a particular roll angle, as well as for transition from one stabilized orientation to another. This redundancy is most likely related to the polyfunctional role of individual RS neurons, which can transmit not only the commands for postural stabilization in different planes but also the commands for steering, for activation of locomotor networks, and the like (Grillner et al., 1995). Involvement of numerous RS neurons in the transmission of each particular command will perhaps promote decoding of the supraspinal messages by the spinal cord.

9. *Coding of spatial information by orientation-dependent characteristics of the response to movement.* Typically, information about spatial orientation of the animal is coded by the orientation-dependent discharge frequency in the corresponding neurons. Experiments on the lamprey have shown that spatial information can also be coded by the orientation-dependent characteristics of the response to movement. Moreover, postural stabilization in our experiments with the artificial feedback was exclusively based on the dynamic, orientation-dependent vestibular responses in RS

neurons (figure 4.16, b-d). This finding strongly suggests that any small head movements during postural stabilization, such as head oscillations caused by periodic locomotor movements, can improve the operation of the postural control mechanisms.

Acknowledgments

This work was supported by grants from the National Institute of Health (NS38022), the Swedish Medical Research Council (11554), the Swedish Society for Medical Research, and the Royal Swedish Academy of Science.

References

Alkon, D.L. (1975) Responses of hair cells to statocyst rotation. *J Gen Physiol* 66: 507-530.

Arshavsky, Y.I., Beloozerova, I.N., Orlovsky, G.N., Panchin, Y.V., and Pavlova, G.A. (1985) Control of locomotion in marine mollusc *Clione limacina:* 1. Efferent activity during actual and fictitious swimming. *Exp Brain Res* 58: 255-262.

Arshavsky, Y.I., Deliagina, T.G., Gamkrelidze, G.N., Orlovsky, G.N., Panchin, Y.V., and Popova, L.B. (1993) Pharmacologically induced elements of the hunting and feeding behavior in the pteropod mollusk *Clione limacina:* 1. Effects of physostigmine. *J Neurophysiol* 69: 522-532.

Arshavsky, Y.I, Deliagina, T.G., and Orlovsky, G.N. (1997) Pattern generation. *Curr Opin Neurobiol* 7: 781-789.

Arshavsky Y.I., Deliagina, T.G., Orlovsky, G.N., Panchin, Y.V., Pavlova, G.A., and Popova, L.B. (1991) Locomotion of *Clione limacina* in relation to various types of behaviour. In Sakharov, D.A., and Winlow, W. (Eds.), *Studies in neuroscience,* pp. 290-315. Manchester and New York: Manchester Univ. Press.

de Burlet, H.M., and Versteegh, C. (1930) Uber Ban und Funktion des Petromyzonlabyrinthes. *Acta Otolaringol* (Suppl. 13): 5-58.

Burt, A., and Flohr, H. (1991) Role of the visual input in recovery of function following unilateral vestibular lesion in the goldfish: I. Short-term behavioural changes. *Behav Brain Res* 42: 201-211.

Deliagina, T.G. (1997a) Vestibular compensation in lampreys: Impairment and recovery of equilibrium control during locomotion. *J Exp Biol* 200: 1459-1471.

Deliagina, T.G. (1997b) Vestibular compensation in lampreys: Role of vision at different stages of recovery of equilibrium control. *J Exp Biol* 200: 2957-2967.

Deliagina, T.G., Arshavsky, Y.I., and Orlovsky, G.N. (1998a) Control of spatial orientation in a mollusc. *Nature* 393: 172-175.

Deliagina, T.G., and Fagerstedt, P. (2000) Responses of reticulospinal neurons in intact lamprey to vestibular and visual inputs. *J Neurophysiol* 83: 864-878.

Deliagina, T.G., Grillner, S., Orlovsky, G.N., and Ullen, F. (1993) Visual input affects the response to roll in reticulospinal neurons of the lamprey. *Exp Brain Res* 95: 421-428.

Deliagina T.G., and Orlovsky, G.N. (1990) Control of locomotion in the freshwater snail *Planorbis corneus:* 1. Locomotory repertoire of the snail. *J Exp Biol* 152: 389-404.

Deliagina, T.G., Orlovsky, Y.I., and Arshavsky, Y.I. (1998b) Control of body orientation in *Clione limacina:* Spatial zones of activity of different neuron groups. *Soc Neurosci Abstr* 840.15.

Deliagina, T.G., Orlovsky, G.N., Grillner, S., and Wallen, P. (1992a) Vestibular control of swimming in lamprey: 2. Characteristics of spatial sensitivity of reticulospinal neurons. *Exp Brain Res* 90: 489-498.

Deliagina, T.G., Orlovsky, G.N., Grillner, S., and Wallen, P. (1992b) Vestibular control of swimming in lamprey: 3. Activity of vestibular afferents. Convergence of vestibular inputs on reticulospinal neurons. *Exp Brain Res* 90: 499-507.

Deliagina, T.G., Orlovsky, G.N., Selverston, A., and Arshavsky, Y. (1999) Neuronal mechanisms for the control of body orientation in *Clione*. 1. Spatial zones of activity of different neuron groups. *J Neurophysiol* 82: 687-699.

Deliagina, T.G., Popova, L.B., and Grant G. (1997) The role of tonic vestibular input for postural control in rats. *Arch Ital Biol* 135: 239-261.

Deliagina, T.G., Zelenin, P.V., Fagerstedt, P., Grillner, S., and Orlovsky, G.N. (2000) Activity of reticulospinal neurons during locomotion in the freely behaving lamprey. *J Neurophysiol* 83: 853-863.

Dieringer, N. (1995) "Vestibular compensation": Neural plasticity and its relations to functional recovery after labyrinthine lesions in frogs and other vertebrates. *Progress in Neurobiology* 46: 97-129.

Drew, T. (1991) Visuomotor coordination in locomotion. *Curr Opin Neurobiol* 1, 652-657.

Feldman, A.G. (1986) Once more on the equilibrium-point hypothesis (lambda-model) for motor control. *J Motor Behav* 18: 17-54.

Fung, J., and Macpherson, J.M. (1995) Determinants of postural orientation in quadrupedal stance. *J Neurosci* 15(2): 1121-1131.

Ghez, C. (1991) Posture. In Kandel, E.R., Schwartz, J.H., and Jessell, T.M. (Eds.), *Principles in neural science*, pp. 596-607. New York: Elsevier.

Grillner, S., Deliagina, T., Ekeberg, Ö., El Manira, A., Hill, R., Lansner, A., Orlovsky G., and Wallen, P. (1995) Neural networks controlling locomotion and body orientation in lamprey. *TINS* 18: 270-279.

Harris-Warrick, R.M., and Marder, E. (1991) Modulation of neural networks for behavior. *Ann Rev Neurosci* 14: 39-57.

Hirschfeld, H., and Forssberg, H. (1991) Phase-dependent modulations of anticipatory postural activity during human locomotion. *J Neurophysiol* 66: 12-19.

Horak, F.B., and Macpherson, J.M. (1995) Postural orientation and equilibrium. In Integration of motor, circulatory, respiratory and metabolic control during exercise, Section 12, Shepard, J., and Rowell, L. (Eds.), *Handbook of physiology*, pp. 1-39. New York: Oxford University Press.

Horak, F.B., and Nashner, L.M. (1986) Central programming of postural movements: Adaptation to altered support-surface configurations. *J Neurophysiol* 55: 1369-1381.

Janse, C. (1982) Sensory system involved in gravity orientation in pulmonate snail *Lymnaea stagnalis*. *J Comp Physiol A* 145: 311-319.

Kably, B., and Dew, T. (1998) Corticoreticular pathways in the cat: II. Discharge activity of neurons in area 4 during voluntary gait modifications. *J Neurophysiol* 80: 406-424.

Kappers, A.C.U., Huber, G.C., and Crosby, E. (1936) The comparative anatomy of the nervous system of vertebrates, including man. New York: MacMillan.

Kasicki, S., Grillner, S., Ohta, Y., Dubuc, R., and Brodin, L. (1989) Phasic modulation of reticulospinal neurones during fictive locomotion and other types of spinal motor activity in lamprey. *Brain Res* 484: 203-216.

Macpherson J., Deliagina, T.G., and Orlovsky, G.N. (1997) Control of body orientation and equilibrium in vertebrates. In Stuart, D., and Stein, P. (Eds.), *Neurons, networks, and motor behavior*, pp. 257-267. Cambridge, MA: MIT Press.

Magnus R. (1924) *Körperstellung*. Berlin: Springer.

Marder, E., and Calabrese, R.L. (1996) Principles of rhythmic motor pattern generation. *Physiol Rev* 76: 687-717.

Massion, J. (1991) Movement, posture and equilibrium: Interaction and coordination. *Prog Neurobiol* 38: 35-56.

Massion, J. (1994) Postural control system. *Curr Opin Neurobiol* 4: 877-887.

Massion, J. (1998) Postural control systems in developmental perspective. *Neurosci Behav Rev* 22: 467-472.

Massion, J., and Dufossé, M. (1988) Coordination between posture and movement: Why and how? *News in Physiol Society* 3: 88-93.

Nashner, L.M., and Forssberg, H. (1986) Phase-dependent organization of postural adjustments associated with arm movements while walking. *J Neurophysiol* 55: 1382-1394.

Nieuwenhuys, R. (1972) Topological analysis of the brain stem of the lamprey *Lampetra fluviatilis. J Comp Neurol* 145: 165-177.

Ohta, Y., and Grillner, S. (1989) Monosynaptic excitatory amino acid transmission from the posterior rhombencephalic reticular nucleus to spinal neurons involved in the control of locomotion in lamprey. *J Neurophysiol* 62: 1079-1089.

Orlovsky, G.N. (1972) The effect of different descending systems on flexor and extensor activity during locomotion. *Brain Res* 40: 359-371.

Orlovsky, G.N. (1991) Gravistatic postural control in simpler systems. *Curr Opin Neurobiol* 1: 621-627.

Orlovsky, G.N., Deliagina, T.G., and Wallén, P. (1992) Vestibular control of swimming in lamprey: 1. Responses of reticulospinal neurons to roll and pitch. *Exp Brain Res* 90: 479-488.

Panchin, Y.V., Arshavsky, Y.I., Deliagina, T.G., Popova, L.B., and Orlovsky, G.N. (1995a) Control of locomotion in marine mollusk *Clione limacina:* VIII. Cerebropedal neurons. *J Neurophysiol* 73: 1912-1923.

Panchin, Y.V., Arshavsky, Y.I., Deliagina, T.G., Popova, L.B., and Orlovsky, G.N. (1995b) Control of locomotion in marine mollusk *Clione limacina:* IX. Neuronal mechanisms of spatial orientation. *J Neurophysiol* 73: 1924-1937.

Rouse, D.T., and McClellan, A.D. (1997) Descending propriospinal neurons in normal and spinal cord-transected lamprey. *Exp Neurol* 146: 113-124.

Rovainen, C.M. (1979) Electrophysiology of vestibulospinal and vestibuloreticulospinal systems in lampreys. *J Neurophysiol* 42: 745-766.

Satterlie, R.A., LaBarbera, M., and Spencer A.N. (1985) Swimming in the pteropod mollusc *Clione limacina:* 1. Behavior and morphology. *J Exp Biol* 116: 189-204.

Selverston, A.I. (1993) Neuromodulatory control of rhythmic behaviors in invertebrates. *Int Rev Cytol* 147: 1-24.

Selverston, A.I. (1995) Modulation of circuits underlying rhythmic behaviors. *J Comp Physiol A* 176: 139-147.

Selverston, A.I., Panchin Y.V., Arshavsky, Y.I., and Orlovsky, G.N. (1997) Shared features of invertebrate central pattern generators. In Stein, P.S.G., Grillner, S., Selverston, A.I., and Stuart, D.G. (Eds.), *Neurons, networks, and motor behavior,* pp. 105-117. Cambridge, MA: MIT Press.

Smith, P.F., and Curthoys, I.S. (1989) Mechanisms of recovery following unilateral labyrinthectomy: A review. *Brain Res Rev* 14: 155-180.

Tsirulis, T.P. (1974) The fine structure of the statocyst in the pteropod mollusk *Clione limacina. J Evolut Biochem Physiol* 10: 181-188.

Ullén, F., Deliagina, T.G., Orlovsky, G.N., and Grillner, S. (1995a) Spatial orientation of lamprey: 1. Control of pitch and roll. *J Exp Biol* 198: 665-673.

Ullén, F., Deliagina, T.G., Orlovsky, G.N., and Grillner S. (1995b). Spatial orientation of lamprey. 2. Visual influence on orientation during locomotion and in the attached state. *J Exp Biol* 198: 675-681.

Viana di Prisco, G., Wallén, P., and Grillner, S. (1990) Synaptic effects of intraspinal stretch receptor neurons mediating movement-related feedback during locomotion. *Brain Res* 530: 161-166.

von Holst, E. (1935) Über den Lichtrückenreflex bei Fischen. *Pubbl Staz Zool Napoli* 15: 143-158.

Wannier, T., Deliagina, T.G., Orlovsky, G.N., and Grillner, S. (1998) Differential effects of reticulospinal system on locomotion in lamprey. *J Neurophysiol* 80: 103-112.

Zelenin, P.V., Deliagina, T.G., Grillner, S., and Orlovsky, G.N. (2000) Postural control in the lamprey—a study with neuro-mechanical model. *J Neurophysiol* 84: 2880-2887.

Zelenin, P.V., Grillner, S., Orlovsky, G.N., and Deliagina, T.G. (2001) Heterogeneity of the population of command neurons in the lamprey. *J Neurosci* 21: 7793-7803.

Development of Balance Control in Typically Developing Children and Children With Cerebral Palsy

Contributions and Constraints of Musculoskeletal Versus Nervous Subsystems

Marjorie Hines Woollacott

Department of Exercise and Movement Science
and Institute of Neuroscience, University of Oregon

When we observe an infant taking her first steps or making that first accurate reach for a bright ball that is rolling across the floor, we may marvel at the development of her walking or reaching abilities, knowing the limitations of those abilities at birth. Early research studying the development of walking and reaching focused on the skill itself in trying to understand the factors contributing to skill development.

However, it has become clear in recent years that the appearance of these skills is really the culmination of the development of many underlying systems, all of which are critical to the emergence of walking and reaching behaviors. One critical system contributing to their development is the postural control system.

A set of observations in the early 1980s by Amiel-Tison and Grenier (1980) gave interesting evidence to support the hypothesis that postural control is critical to motor skill development. They observed that when the chaotic movements of the head that typically disturb the infant's seated balance were stabilized, behaviors and movements typically seen in more mature infants emerged. For example, when they stabilized the head of newborn infants, they noted that the infants began to attend to the clinician, reach for objects, and maintain their arms at their sides, with the fingers open, suggesting inhibition of the grasp and Moro reflexes that would normally be present in an infant of this age. These observations suggest that postural control of the head and trunk is a critical component of reaching in infants and that it may be the development of this system, rather than of manipulation per se, that is the critical factor constraining the emergence of accurate reaching during the first 4 to 5 months of life.

Similar observations related to the development of locomotion have also been reported. For example, when supported under the arms and tipped slightly forward while on a treadmill, newborn infants begin to take rhythmic steps, showing that the nervous system circuitry for locomotor behavior is present at this age (Forssberg, 1985; Prechtl, 1997; Thelen et al., 1989). Although this behavior has been reported to disappear at about 2 months of age, suggesting that the circuitry for this behavior is suppressed, it has also been shown that submersing the infant in water, thus countering the effects of gravity, allows the behavior to become stronger during this time period. This suggests that it is an increase in the infant's weight, rather than suppression of the neural circuitry, that is the constraint on the ability to observe this behavior in many infants between 2 months of age and the onset of self-initiated walking many months later (Thelen et al., 1989).

It is well-known that children with cerebral palsy have difficulties with voluntary skills such as walking and reaching, in addition to underlying impairments in balance control. Studies have documented both delays in the onset of independent stance and walking in many children and an inability to balance sufficiently to stand and walk in others, depending on the severity of the symptoms of cerebral palsy (Crothers and Paine, 1988). In addition, recent studies have shown that adequate postural alignment and stability while sitting improve the functional performance of children with cerebral palsy during the performance of manipulation skills (McClenaghan et al., 1992; Seeger et al., 1984).

These studies on both typically developing children and children with cerebral palsy support the concept that an immature or dysfunctional postural system is a limiting factor or a constraint on the emergence of other behaviors such as coordinated arm and hand movements and walking skills. Thus, to understand the emergence of mobility and manipulatory skills in children, researchers need to understand the postural substrate for these skills. Similarly, understanding the best therapeutic approach for a child with cerebral palsy who has difficulties in walking or reaching skills requires the knowledge of any limitations in their postural abilities.

Although posture or balance has traditionally been discussed as if it were controlled by a single system, research during the last 15-20 years has shown that a variety of both nervous and musculoskeletal subsystems contribute to balance control. Neural subsystems contributing to balance control include the sensory systems (particularly the visual, somatosensory, and vestibular systems), motor systems (both motor cortex and other descending systems contributing to muscle response coordination and force generation), and higher level systems such as the cerebellum and basal ganglia, which contribute to modulation of responses under changing task and environmental conditions. In addition, musculoskeletal components such as skeletal alignment and muscle strength are key contributing factors to successful balance control and development.

Several neural factors have been hypothesized to contribute to constraints on balance control in children with cerebral palsy, including spasticity, or hyperactive stretch reflexes, and problems with muscle coordination, including poorly organized postural responses and increased coactivation of muscles at individual joints. In addition, children are known to have musculoskeletal constraints, including muscle weakness and a crouched posture, both of which could also contribute to balance problems.

Development of Stance Balance Control in Typically Developing Children

In the following pages, we will first review some of the research performed in the last 20 years on the contributions of the different neural and musculoskeletal systems to balance control in typically developing children. We will then explore studies that have sought to determine the extent to which dysfunction or delays in specific neural or musculoskeletal subsystems contribute to balance difficulties in children with cerebral palsy. Early research on balance constraints in children with cerebral palsy focused primarily on reflexly based neural factors such as spasticity. More recent work has begun to focus on the contributions of other neural systems and on the musculoskeletal system (Brogren et al., 1996; Burtner et al., 1998; Nashner et al., 1983).

Motor Coordination

Balance control can be divided into (1) static or steady-state control, which contributes to the control of quiet stance balance, and dynamic balance control, including both (2) reactive balance control (e.g., responses to unexpected balance threats) and (3) proactive or anticipatory balance control, involving the generation of balance responses in advance of a destabilizing condition. Responses in the third category could include activating postural muscles in advance of making a reaching movement so as to prevent a balance disturbance caused by the destabilizing forces due to the center-of-mass (COM) changes during the movement. In the following section, we will focus on research relating to the development of dynamic balance control, including both reactive and proactive balance.

A research approach that was created to study reactive balance control (Nashner, 1976, 1977) and was subsequently used to study the development of balance control in the infant (Forssberg and Nashner, 1982; Sveistrup and Woollacott, 1996) is a hydraulically activated platform that moves horizontally to create external threats to balance. These threats are similar to those produced when a bus on which one is standing starts to move. A second approach to the study of reactive balance uses accelerations or decelerations of a treadmill on which the subject is standing or walking to threaten balance and activate balance adjustments (Berger et al., 1985).

Neurophysiological Approach: Electromyography

Most studies on the development and refinement of stance balance control have been cross-sectional and have examined children who already stand independently, in the age range of 1-10 years. For example, studies by Forssberg and Nashner (1982) and Shumway-Cook and Woollacott (1985) showed that compensatory responses to postural perturbations were present in newly walking children as young as 15 months of age. These responses showed an organization like that seen in adults, with ankle muscles closest to the base of support being activated first at about 100-110 ms, followed by more proximal thigh muscles about 20-40 ms later. Thus, when the platform moved backward, causing forward sway, the gastrocnemius was activated first, followed by the hamstring muscles. Although responses were organized in a distal-to-proximal manner, as in adults, response onsets were delayed and the responses were larger in amplitude, longer in duration, and more variable than those of older children and adults. The youngest children (1-3 years) also showed strong coactivation of antagonist muscles along with the agonist leg muscles, indicating a strategy of stiffening the joints when responding to a balance threat, perhaps to reduce the degrees of freedom to be controlled. These children also showed increased sway magnitudes and more oscillatory behavior in their recovery from balance threats than did older children.

Research by Berger et al. (1985), in which children were given perturbations while standing or walking on a treadmill, compared the prevalence of monosynaptic stretch reflexes versus longer latency automatic postural responses in children of similar ages. They noted that 1-year-olds showed large monosynaptic stretch reflexes followed by long-lasting, longer latency postural responses. They found that the monosynaptic reflexes were reduced in older children and disappeared after about 4 years of age. In addition, the longer latency responses showed reduced burst durations in the older children, indicating more efficient balance responses.

Biomechanical Approach: Muscle Torques

In addition to using neurophysiological tools to study the development of reactive balance control, one can use biomechanical tools such as kinetic analysis. In this case, one can measure the forces at the ankle, knee, and hip joints used to recover from balance threats and compare these force characteristics in children of increasing developmental levels. One can also measure center of pressure (COP) trajectories used to recover from balance threats and the time it takes to stabilize COP as the infant returns to her quiet stance position.

Recent work by Roncesvalles et al. (2001) and Sundermier et al. (2001) used kinetic analysis to examine the development of reactive balance control in infants from 9 months to 10 years of age. To compare similar sizes of balance threats across different age groups, perturbation magnitudes were normalized to 41% (±7.4) of foot length with theoretical angular velocity induced by the perturbation set at 0.38 rad/s (±0.03).

First, COP trajectories that occurred in response to the platform displacement were analyzed. These were first passive movements of the COP as the child's feet were displaced backward by the platform movement and then active movements of the COP during the child's recovery from the balance threat, as the body was re-aligned over the feet. In examining COP trajectories used to recover from balance threats, it was noted that children just learning to stand and walk (1-year-olds) showed substantially longer COP paths compared to older children. The younger children were also the slowest to recover stability, taking about 2 s to return to quiet stance. The older children (7- to 10-year-olds) were able to recover stability in about half the time (i.e., 1.1 s).

To determine what factors contributed to this improved timing and efficiency in balance recovery, the researchers examined the torque profiles from the ankle, knee, and hip generated during balance recovery for children of different age groups. They found that, in contrast to older children and adults, who rapidly generated large torques at each of the joints, younger children (standers and walkers: 9-23 months of age) used multiple torque adjustments before regaining control. Figure 5.1 shows torque profiles of children from 9-13 months (new standers), 14-23 months (advanced walkers), 2-3 years (runners/jumpers), 4-6 years (gallopers), and 7-10 years (skippers). Note that there are at least three bursts of torque production at the ankle, knee, and hip in the standers and walkers, whereas this is reduced to two and then one burst in the older age groups. In addition, the youngest age groups tended to over- and undershoot torque requirements in attempting to regain stability, with many torque reversals being apparent.

Although this study does not provide a window into the forces created by individual muscles at a joint, it allows us to see the sum of the activity of all the muscles being activated, which electromyographic (EMG) recording is unable to do. This study, in combination with previous work studying the refinement of balance strategies using neurophysiological and kinematic measures (Forssberg and Nashner, 1982; Shumway-Cook and Woollacott, 1985), allows us to formulate the following general developmental principles underlying the refinement of postural control in

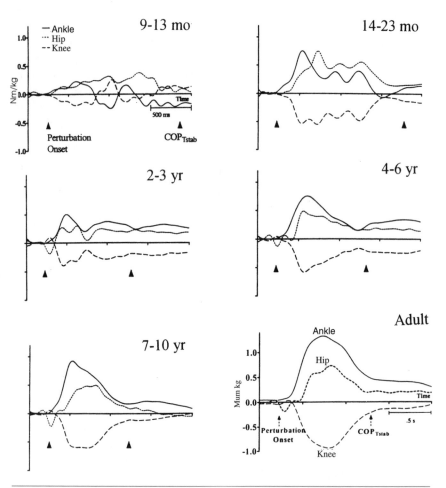

Figure 5.1 Muscle torque profiles (ankle, knee, and hip) from children of 9-13 months (new standers), 14-23 months (advanced walkers), 2-3 years (runners/jumpers), 4-6 years (gallopers), and 7-10 years (skippers) as they recovered balance after experiencing a backward platform perturbation. Onset of perturbation and time of recovery of balance are marked by arrows.

Reprinted from Roncesvalles, Woollacott, and Jensen 2001.

children from 1-10 years of age: (1) both response onsets and total recovery time show a significant reduction with increased development and experience; (2) both sway and COP excursions are significantly reduced, with a tendency to overshoot and undershoot their COP original resting range being eliminated as children develop; (3) responses move from being multimodal, with multiple torque bursts at a joint, to unimodal as children develop balance skills; and (4) responses are essentially mature by 7-10 years of age.

Other work by Assaiante and Amblard (1995) has used kinematic analysis to explore changes in the control of body segments during balance control while walking in children between 1 and 10 years of age. They found that until about 6 years of age, children control walking balance in what they call a "bottom-up" mode, using the support surface for reference. This is similar to the results found by Shumway-Cook and Woollacott (1985) and Forssberg and Nashner (1982), showing that young children use muscle patterns during stance balance control that are organized using the support surface for reference. In this case, reactive balance responses began with the muscles closest to the base of support and radiated upward.

The work mentioned previously has provided knowledge about the refinement of balance control in children who are already standing and walking independently. But when do balance responses actually develop in order to allow the child to take her first independent steps?

To answer this question, Sveistrup and Woollacott (1996) performed a longitudinal study on nine children from 33 to 65 weeks of age (developmental levels of pre-pull-to-stand behavior through 3 months of independent walking), examining the emergence of muscle response patterns elicited by unexpected threats to balance. Infants stood on a movable platform similar to those used in previous studies, with a parent sitting in front of the infant. If the infant could not stand independently, a toy castle was placed in front of the infant to be used for support.

The researchers found a clear developmental progression in the emergence of muscle responses for children from early pull to stand (33 weeks) through 3 months' experience in independent walking (65 weeks). For children in the pre-pull-to-stand stage of development, there were no organized patterns of EMG activation in response to balance threats. Muscle responses were typically absent or present only in isolated muscles, with large amounts of tonic background activity present. By pull to stand, phasic bursts were more clearly identifiable above the tonic background activity, with some infants showing responses in multiple muscles on at least one trial. However, there was still high variability from trial to trial. By independent stance, the tonic activity had further decreased and all infants responded to at least one platform perturbation with all three muscles, in an ascending sequence starting with the stretched ankle muscle. Through independent walking and late independent walking (3 months' experience), clear phasic responses continued to be present, with a higher probability of seeing all three muscles responding on each trial and lower levels of background activity. Figure 5.2 shows the average probability of seeing one, two, or three muscles activated in the children at each developmental level. Note that the probability of seeing a three-muscle response was very low in early pull to stand but increased to high levels by the time children were walking for 2-3 months.

In studies on the development of locomotion, Bril and Breniere (1993) show a two-phase process, the first of which appears to involve learning to control balance (the first 3-5 months of independent walking) and the second of which involves the progressive refinement of the locomotor pattern (5 months' to about 4 years' experience). In the first 3-4 months, they have shown a rapid decrease in the double-

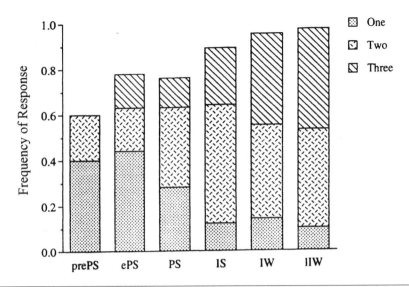

Figure 5.2   The average probability of seeing one, two, or three muscles activated in response to a balance threat for the children at each developmental level (PS = pull to stand, IS = independent stance, IW = new walkers, IIW = experienced walkers).
Reprinted from Sveistrup and Woollacott 1996.

support phase of gait, an increase in step length, and a decrease in step width, all of which relate to the mastery of balance control. These results probably relate to this first period in balance development explored by Sveistrup and Woollacott (1996), with their second period relating to the refinement of balance control that occurs between 1 and 7 years of age (Forssberg and Nashner, 1982; Roncesvalles et al., 2001; Shumway-Cook and Woollacott, 1985).

Adaptation to Changing Magnitudes of Balance Threats

The previous studies have focused on the development of motor coordination underlying balance control in a relatively stereotyped situation. However, children typically balance under a variety of conditions, with balance threats varying from those in which the center of mass stays well within the base of support, requiring only a small amount of force for recovery, to large threats that move the center of mass outside the base of support and require one or more steps to remain upright. How do children develop the ability to respond to these varying balance task demands?

A second study by Roncesvalles et al. (2000) examined the development of compensatory stepping skills in children (9-19 months of age) during the first 6 months of independent stance and walking. The following four behavioral groups were included: standers, new walkers (<2 weeks' walking experience), intermediate walkers (1-3 months' walking experience), and advanced walkers (3-6 months' walking experience). In these experiments, the standing balance of the infants was perturbed using increasingly faster support surface translations (8 cm [3.15 in.] at 15, 20, or 25 cm/s [5.9, 7.9, or 9.8 in./s]). Even at the slower perturbations (15 cm/s), all of the standers collapsed on almost every trial while the new walkers fell 62% of the time and used feet-in-place responses on the other 38% of the trials, with no ability to perform steps. The intermediate walkers were the first to show stepping responses (in 29% of the trials), with feet-in-place responses in 65% of the trials and only 6% falls. All advanced walkers showed feet-in-place responses in 100% of the trials (see figure 5.3).

As perturbation velocities increased, both new and intermediate walkers showed a higher probability of falls, as can be seen in figure 5.3. New walkers showed almost no stepping responses, while the intermediate walkers showed the ability to respond to these increasing threats to balance by increasing step frequency from 29% to 44% and finally to 53% as velocity increased. The advanced walkers also showed mainly feet-in-place responses but were able to shift to a stepping strategy with the higher velocities as well (0%, 29%, and 33% stepping probability for the three velocities).

It is interesting to note that the onset latency for initiating a lateral COP shift to begin the step was more delayed in the children with increased walking experience, indicating higher levels of stability. In addition, the COP shift was largely mediolateral in the new walkers, with intermediate and advanced walkers showing significantly smaller lateral excursions and larger anteroposterior excursions than the new walkers.

These data suggest that the ability to adapt balance responses to increasing balance threats is not available to the new standers and new walkers, with almost no children able to take a step to recover balance. It begins to develop in infants with 1-3 months of walking experience and is relatively refined by 6 months of walking experience, with a stepping strategy readily available and little mediolateral insta-bility in these children when taking steps (Roncesvalles et al., 2000).

Sensory Contributions

As mentioned earlier, three different sensory systems contribute to balance control in the adult: the visual, somatosensory, and vestibular systems. Depending on the type of balance task a person is performing and the environment a person is in, different senses tend to dominate in the control of balance. For example, during quiet stance, all three systems contribute to balance control (Diener et al., 1986). Lee and Lishman (1975) have also shown that when adults are asked to balance on a narrow beam, visual control of balance is increased. However, somatosensory inputs are the major contributors to reactive balance control under normal conditions (Dietz et al., 1991).

Are there specific sensory systems that dominate balance control during different periods in development? Work by Foster et al. (1996), Lee and Aronson (1974), Sundermier and Woollacott (1998), and others suggests that vision may be a primary contributor to both quiet stance and reactive balance during the transition to inde-pendent stance. Its influence then appears to decline, with somatosensory inputs dominating in reactive balance control for older children and adults. Lee and Aronson

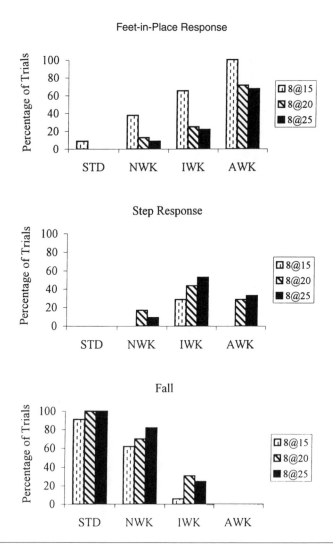

Figure 5.3   Percentage of trials in which children used either feet-in-place responses, stepping responses, or fell in response to platform displacements of 8 cm (3.15 in.) at 15, 20, and 25 cm/s (5.9, 7.9, or 9.8 in./s). Abbreviations: STD: standers; NWK: new walkers; IWK: intermediate walkers; AWK: advanced walkers.

Reprinted from Roncesvalles, Woollacott, and Jensen 2000.

(1974) showed that visual flow information simulating body sway during quiet stance caused new walkers to sway, stagger, or fall; however, children aged 2 years or older were minimally affected by the information. Foster et al. (1996) showed that in infants as young as 5 months of age, during supported standing, visual flow information simulating body sway activates well-organized responses in the muscles of the legs and trunk. It is interesting to note that leg muscle postural responses are well organized in supported infants as early as 5 months of age in response to visual flow information simulating sway, whereas support surface displacements do not produce well-organized responses in standing infants until they are about 9-11 months of age and transitioning to independent stance.

To determine if there are also significant contributions of vision to automatic postural responses to support surface displacements in infants, Sundermier and Woollacott (1998) tested two groups of infants, prewalkers and early walkers, on a movable platform under eyes-open and eyes-closed conditions (a peekaboo game was played with the infants in which an opaque scarf was put over their eyes for the perturbation, then quickly removed). They found that there was no measurable effect of visual condition on muscle response characteristics for the prewalkers. However, for the new walkers, the integrated gastrocnemius activity (the response magnitude) increased significantly in eyes-open versus eyes-closed conditions (p <.05). Thus, it appears that in the transition to independent walking, visual cues contribute to the faster, more functionally appropriate automatic postural response pathways, in addition to postural control during quiet stance. As children gain experience in independent stance, somatosensory inputs are hypothesized to become the primary contributors to balance control, as found in young adults.

Musculoskeletal Contributions

It has been suggested that a primary rate-limiting factor for the emergence of independent walking is the development of sufficient muscle strength to support the body during static balance and walking (Thelen and Fisher, 1982). Although it is difficult to measure maximal voluntary contraction in the leg muscles of infants, other experimental methods have been used to estimate the strength of the leg muscles. For example, Roncesvalles and Jensen (1993) have measured the muscle forces produced by infants of 4 to 8 months of age as they bounced up and down. They found that by 6 months of age, infants were producing forces well beyond their own body weight. These results suggest that infants have the strength to support their body weight against the force of gravity in the standing position well before the emergence of independent stance. Thus, muscle strength is probably not a primary constraint on the emergence of stance postural control in these children.

Proactive (Anticipatory) Postural Development

Do infants develop proactive postural control in parallel with reactive postural control? Studies have shown that infants who can sit independently (age 9 months) show activation of the postural muscles of the trunk in advance of most but not all reaching movements. In one study, infants of 9 months of age were balanced on the thigh of their parent, with support only at the hip, while toys were presented to them (Hofsten and Woollacott, 1989). During most reaching movements, they activated postural muscles of the trunk in advance of arm muscles. This allowed them to stabilize the trunk in preparation for the forward movement of the center of mass resulting from the forward reach for the toy. Studies on proactive balance control during stance (Woollacott and Shumway-Cook, 1986) have shown that by 4 to 6 years of age, children activate leg muscle responses in advance of arm muscles to stabilize the body before a pulling or pushing movement.

Development of Stance Balance Control in Children With Cerebral Palsy

The research mentioned earlier on typically developing children suggests that constraints on balance development relate chiefly to the ability to organize effective muscle responses to external threats to stability. Vision appears to play a role in balance control during the transition to independent stance, with visual contributions diminishing with increased experience in standing and walking. Muscle strength appears to be adequately developed before the transition to independent stance and thus is not a constraint on balance development in these children.

In the following section, we will examine the extent to which dysfunction or delays in the development of specific neural or musculoskeletal subsystems contribute to balance difficulties in children with cerebral palsy.

Motor Coordination

Constraints on the development of muscle activation patterns used during both reactive and proactive balance control have been studied in children with cerebral palsy using a paradigm identical to that described previously for typically developing children (Burtner et al., 1998; Nashner et al., 1983; Woollacott et al., 1998). In one of the first studies using a systems approach to examine the constraints on development of reactive balance control in children with cerebral palsy, Nashner et al. (1983) compared responses of children with different types of cerebral palsy (spastic hemiplegia, spastic diplegia, and ataxia) to those of age-matched control children. They found that the children showed consistent motor coordination deficits. For example, children with spastic hemiplegia showed both delayed muscle response onsets and disorganized activation patterns in the muscles of the hemiplegic leg. Thus, in response to backward platform movements causing forward sway, the hamstring muscles were activated before the gastrocnemius muscles, giving a proximal-to-distal muscle response sequence rather than the distal-to-proximal sequence seen in typically developing children.

How did this affect balance? The forces created by the hemiplegic muscles of the involved leg were of low level and therefore out of balance with those from the less involved leg. This created an inefficient balance recovery due to mediolateral oscillations. It is interesting that the clinical exam showed spastic gastrocnemius responses, yet the postural tests showed delayed, smaller gastrocnemius responses. These results suggest that clinical exams testing reflexes may not predict performance on active postural tasks.

Children with ataxia (due to cerebellar lesions) did not show disordered patterns in muscle recruitment; however, they demonstrated longer onset latencies than control children, with trial-to-trial variations noted.

In a recent study, Burtner et al. (1998) compared the muscle response characteristics of children who were in the transition period to independent stance and who were normally developing with children who had spastic diplegic cerebral palsy and were of similar developmental levels. Three groups of children participated: a pull-to-stand group (2 CP, 2 control), a young walker group with 2-4 years of walking experience (2 CP, 2 control), and an experienced walker group with 8-14 years of walking experience (3 CP, 3 control). Since it is known that spasticity is present in the posterior leg muscles of children with cerebral palsy, analysis was focused on the responses of these muscles (gastrocnemius, hamstrings, trunk extensors). Balance threats were created with backward support surface displacements causing forward sway.

Results showed that with increased walking experience, the control children showed a trend toward decreased duration of muscle responses, whereas children with cerebral palsy showed a prolonged duration of muscle activation at all stages of stance and walking experience. The children with cerebral palsy in the transition to independent stance also showed an increased frequency of reversals of muscle response recruitment, with proximal muscles activated before distal muscles, as was shown in the study by Nashner et al. (1983) for older children with more walking experience. When compared to control children, they also showed increased recruitment of antagonist muscles when responding to balance threats and a decreased activation of trunk musculature. The control children in the pull-to-stand stage of development did show some disorganization in their temporal organization of responses, with agonist/antagonist coactivation and reversals seen in a small number of trials. However, with increased walking experience (young walker and experienced walker groups), this disorganization disappeared. Accompanying these less efficient muscle response characteristics were increased times to COP stabilization and increased COP excursions during balance recovery.

Sensory Contributions

As noted earlier, the visual system of typically developing children plays a dominant role in stance postural stability up to 2 years of age, when they have gained experience in independent stance and walking (Foster et al., 1996). Knowing that children with cerebral palsy have sensorimotor deficits and delays in the onset of independent stance, it is possible that they would show visual dominance in postural control longer than typically developing children. In a preliminary study, Bai et al. (1987) looked at visual contributions to balance control in children with cerebral palsy using the "moving room" research paradigm described earlier for studying these contributions to balance control in typically developing children. They compared normative data for 3-year-old children to a 3-year-old child with ataxia. In response to the visual flow condition, the typically developing 3-year-old children attenuated their responses quickly with little sway in response to the room movement after 10 trials. However, the child with ataxic cerebral palsy staggered or fell on almost every trial. Over as many as 30 trials, this child showed little response attenuation to the room movement. Results of this pilot study suggest that the visual system may be the dominant system in controlling balance in children with cerebral palsy, even after 2 years of age. Since the onset of walking is also typically delayed in these children, it is unclear to what extent sensorimotor deficits versus reduced experience in stance balance contribute to this extended period of visual dominance.

In the previous example testing visual contributions to balance control, a sensory conflict condition is created in which visual stimuli create the sensation of body sway while vestibular and somatosensory stimuli do not signal body sway. Additional research has been performed to determine the extent to which children with different types of cerebral palsy have deficits in the ability to adapt to sensory conflict conditions. Nashner et al. (1983) showed that children with ataxia have serious difficulty with sensory organization tasks, whereas children with spastic hemiplegia or diplegia have less difficulty. In this study, spontaneous sway was recorded in six sensory conditions. In the first three conditions, the children stood on a firm surface (normal somatosensory inputs) with either vision present, vision absent, or in a room that rotated with body sway, eliminating sway-related visual inputs (visual conflict). In the second three conditions, the child stood on a support surface that rotated with body sway, eliminating sway-related somatosensory information (sensory conflict), with either vision present, vision absent, or visual inputs unrelated to body sway.

Children with all three types of cerebral palsy had more sway than control children in the first three sensory conditions, with the children with ataxia having the most difficulty, particularly in the visual conflict condition. In the last three somatosensory conflict conditions, all children had more sway than in the first set of conditions; however, no typically developing children lost balance. The children with cerebral palsy were much more unstable, with one child with spastic hemiplegia and two children with ataxia losing their balance in visual and somatosensory conflict conditions (condition 6). The authors thus concluded that children with cerebral palsy (especially those with ataxia) had difficulty dealing with sensory conflict information and thus were unable to organize sensory information effectively for postural control.

Musculoskeletal Contributions

Many children with cerebral palsy are constrained to stand in a crouched posture with increased hip and knee flexion (Gage, 1991), due to spasticity and other musculoskeletal problems such as contractures. These musculoskeletal system constraints may also contribute to problems in organizing muscle responses effectively when recovering from threats to balance. To determine the contributions of a crouched stance posture to deficits in muscle response organization in reactive balance control, Burtner et al. (1998) performed the following study. Typically developing children were asked to stand both in their habitual manner and in a crouched stance, similar to that of a group of age-matched children with cerebral palsy, and their muscle response characteristics during reactive balance control (response to support surface displacements) were recorded. Results showed that during the crouched stance trials, the typically developing children activated muscles more frequently in a proximal-to-distal fashion, showing the reversals frequently seen in children with spastic diplegia. This suggests that musculoskeletal constraints are a significant contributor to muscle response abnormalities in children with spastic diplegic cerebral palsy.

These results are similar to those of Seinko-Thomas et al. (1995), who asked typically developing children to walk in a crouched gait and found that their locomotor muscle activation patterns became more similar to those of children with cerebral palsy, with high levels of agonist and antagonist muscle coactivation.

Constraints on muscle strength may also compromise balance control in children with spastic diplegia or hemiplegia. Wiley and Damiano (1998) have shown that maximum voluntary contraction levels for muscles in the lower extremity are significantly lower for children with cerebral palsy than in age-matched controls. They noted that weakness was more pronounced in the distal muscles. This is important since it is the ankle muscles that normally contribute the most to balance recovery.

Many children with spastic diplegia are fitted with ankle-foot orthoses (AFOs) to reduce equinus problems. Do orthotic devices such as ankle-foot orthoses actually improve muscle response characteristics to balance threats, or do they impair these responses? In a recent study, Burtner et al. (1999) examined postural muscle response characteristics of both typically developing children and children with spastic diplegia when wearing either no AFOs, solid AFOs, or dynamic AFOs. They found that the percentage of trials in which children used the normal ankle strategy (sway primarily at the ankle), activated the gastrocnemius muscle, and used a distal-to-proximal muscle response sequence was reduced when the children wore the solid AFO. The dynamic AFO allowed muscle response characteristics that were closer to those seen with no AFO. Figure 5.4 shows the percentage of trials in which both control children and children with spastic diplegia used ankle strategies in response to balance threats under the three AFO conditions. Note that the largest reduction in ankle strategies to recover balance in both groups was in trials where they were wearing the solid AFO.

Proactive Balance Control

Do children with cerebral palsy have difficulty organizing postural responses in a proactive manner to stabilize the body in advance of center-of-mass displacements caused during reaching or pulling movements? To examine this question, Nashner et al. (1983) recorded both lower extremity and arm muscle responses of typically

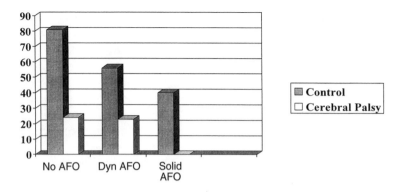

Figure 5.4 Percentage of trials in which children (control children versus children with spastic diplegic cerebral palsy) used ankle strategies (sway primarily at the ankle) to recover from a balance threat (backward platform displacement causing forward sway) under the three AFO conditions (no AFO, dynamic AFO, solid AFO).

developing children and children with cerebral palsy while the standing children were pulling or pushing on a handle. They again compared children with three different types of cerebral palsy (hemiplegia, diplegia, ataxia) to age-matched control children. They found that the control children demonstrated activation of lower extremity muscles in a proactive manner (before the arm muscles) to stabilize the body in advance of the COM displacement caused by the arm movement. However, the children with cerebral palsy typically failed to use proactive recruitment of postural muscles, relying instead on reactive recruitment of these muscles.

Conclusion

Studies examining the development and refinement of balance control in typically developing children support the following general principles of the development of motor coordination:

1. Responses develop gradually, first with one or two muscles responding, and then with three muscles being activated more consistently.

2. Responses develop in a distal-to-proximal activation sequence.

3. Tonic background activity in both agonist and antagonist muscles is reduced as responses become more consistent.

4. Both response onsets and total recovery time show a significant reduction with increased development and experience.

5. Both sway and COP excursions are significantly reduced, with a tendency to overshoot and undershoot their original COP resting range being eliminated as children develop.

6. Responses move from being multimodal, with multiple torque bursts at a joint, to unimodal as children develop balance skills.

7. Responses are essentially mature by 7-10 years of age.

Sensory contributions to balance control appear to be dominated by the visual system during the transition to independent stance, with somatosensory inputs becoming more important as balance experience grows. Muscular strength is adequate for balance, even in pull-to-stand stages of development. Thus, this does not seem to be a rate-limiting factor for balance development.

The development of balance control in children with cerebral palsy shows the following constraints in motor coordination mechanisms:

1. Muscle response onsets are delayed, with burst durations being longer than in typically developing children.

2. A proximal-to-distal response organization is more common.

3. More coactivation of antagonist muscles is used.

4. Sway excursions are larger than for typically developing children.

Sensory contributions to balance control are dominated by the visual system, with children showing difficulty in adapting to visual conflict or somatosensory conflict conditions. Muscular strength is significantly weaker in children with spastic diplegia than in typically developing children, especially in the ankle muscles. The musculoskeletal constraints of the crouched posture in which children with spastic diplegia stand also contribute to muscle response inefficiency, including proximal-to-distal response organization and increased antagonist muscle coactivation.

Acknowledgments

This research was supported by National Science Foundation grant No. IBN-9514544 and NIH grant no. AG-05317.

References

Amiel-Tison, C., and Grenier, A. (1980) *Evaluation neurologique du nouveau-ne et du nourrisson* [Neurological evaluation of the human infant], p. 81. New York: Masson.

Assaiante, C., and Amblard, B. (1995) An ontogenetic model for the sensorimotor organizations of balance control in humans. *Human Movement Science* 14: 13-43.

Bai, D.L., Bertenthal, B.I., and Sussman, M.D. (1987) Children's sensitivity to optical flow for the control of stance. *Abstracts of the Third Annual East Coast Clinical Gait Laboratory Conference.* Bethesda, MD: National Institutes of Health.

Berger, W., Quintern, J., and Dietz, V. (1985) Stance and gait perturbations in children: Developmental aspects of compensatory mechanisms. *Electroencephalogr Clin Neurophysiol* 61: 385-395.

Bril, B., and Breniere, Y. (1993) Posture and independent locomotion in childhood: Learning to walk or learning dynamic postural control. In Savelsbergh, G.J.P. (Ed.), *The development of coordination in infancy,* pp. 337-358. Amsterdam: North-Holland.

Brogren, E., Hadders-Algra, M., and Forssberg, H. (1996) Postural control in children with spastic diplegia: Muscle activity during perturbations in sitting. *Developmental Medicine and Child Neurology* 38: 379-388.

Burtner, P., Qualls, C., and Woollacott, M.H. (1998) Muscle activation characteristics of stance balance control in children with spastic diplegia. *Gait and Posture* 8: 163-174.

Burtner, P., Woollacott, M., and Qualls, C. (1999) Stance balance control with orthoses in a group of children with cerebral palsy. *Developmental Medicine and Child Neurology* 41: 748-757.

Crothers, B., and Paine, P.S. (1988) The natural history of cerebral palsy: No. 2. Classics in developmental medicine. London: MacKeith Press.

Diener, H.C., Dichgans, J., Guschlbauer, B., and Bacher, M. (1986) Role of visual and static vestibular influences on dynamic posture control. *Human Neurobiology* 5: 105-113.

Dietz, V., Trippel, M., and Horstmann, G.A. (1991) Significance of proprioceptive and vestibulo-spinal reflexes in the control of stance and gait. In Patla, A.E. (Ed.), *Adaptability of human gait,* pp. 37-52. Elsevier: Amsterdam.

Forssberg, H. (1985) Ontogeny of human locomotor control: I. Infant stepping, supported locomotion, and transition to independent locomotion. *Exp Brain Res* 57: 480-493.

Forssberg, H., and Nashner, L. (1982) Ontogenetic development of postural control in man: Adaptation to altered support and visual conditions during stance. *J Neurosci* 2: 545-552.

Foster, E., Sveistrup, H., and Woollacott, M.H. (1996) Transitions in visual proprioception: A cross-sectional developmental study of the effect of visual flow on postural control. *J Motor Behav* 28: 101-112.

Gage, J.R. (1991) *Gait analysis in cerebral palsy.* Oxford: MacKeith Press.

Hofsten, C. von, and Woollacott, M. (1989) Anticipatory postural adjustments during infant reaching. *Neurosci Abstr* 15: 1199.

Lee, D.N., and Aronson, E. (1974) Visual proprioceptive control of standing in human infants. *Perception and Psychophysics* 15: 529-532.

Lee, D.N., and Lishman, R. (1975) Visual proprioceptive control of stance. *Journal of Human Movement Studies* 1: 87-95.

McClenaghan, B.A., Thombs, L., and Milner, M. (1992) Effects of seat surface inclination on postural stability and function in the upper extremities of children with cerebral palsy. *Developmental Medicine and Child Neurology* 34: 40-48.

Nashner, L. (1976) Adapting reflexes controlling the human posture. *Exp Brain Res* 26: 59-72.

Nashner, L.M. (1977) Fixed patterns of rapid postural responses among leg muscles during stance. *Exp Brain Res* 30: 13-24.

Nashner, L.M., Shumway-Cook, A., and Marin, O. (1983) Stance posture in select groups of children with cerebral palsy: Deficits in sensory organization and muscular coordination. *Exp Brain Res* 49: 393-409.

Prechtl, H.F.R. (1997) The importance of fetal movements. In Connolly, K.J., and Forssberg, H. (Eds.), *Neurophysiology and neuropsychology of motor development,* pp. 42-35. London: MacKeith Press.

Roncesvalles, N.C., and Jensen, J. (1993) The expression of weight-bearing ability in infants between four and seven months of age. *Sport Exerc Psychol* 15: 568.

Roncesvalles, N.C., Woollacott, M., and Jensen, J. (2000) The development of compensatory stepping skills in children. *J Motor Behavior* 32: 100-111.

Roncesvalles, N.C., Woollacott, M., and Jensen, J. (2001) Development of lower extremity kinetics for balance control in infants and young children. *Journal of Motor Behavior* 33: 180-192.

Seeger, B.R., Caudrey, D.J., and O'Mara, N.A. (1984) Hand function in cerebral palsy: The effect of hip-flexion angle. *Dev Med Child Neurol* 26: 601-606.

Shumway-Cook, A., and Woollacott, M. (1985) The growth of stability: Postural control from a developmental perspective. *J Motor Behavior* 17: 131-147.

Sienko-Thomas, S., Moore, C., Kelp-Lenane, C., and Norris, C. (1995) Simulated gait patterns: The resulting effects on gait parameters, dynamic electromyography, joint moments, and physiological cost index. *Gait and Posture* 4: 100-107.

Sundermier, L., and Woollacott, M. (1998) The influence of vision on the automatic postural muscle responses of newly standing and newly walking infants. *Exp Brain Res* 120: 537-540.

Sundermier, L., Woollacott, M., Roncesvalles, N., Jensen, J. (2001) The development of balance control in children: Comparisons of EMG and kinetic variables and chronological and developmental groupings. *Exp Brain Res* DOI 10.1007/S002210000579. Published online 13 Dec.

Sveistrup, H., and Woollacott, M.H. (1996) Longitudinal development of the automatic postural response in infants. *J Motor Behavior* 28: 58-70.

Thelen, E., and Fisher, D.M. (1982) Newborn stepping: An explanation for a "disappearing reflex." *Developmental Psychology* 18: 760-775.

Thelen, E., Ulrich, B.D., and Jensen, J.L. (1989) The developmental origins of locomotion. In Woollacott, M.H., and Shumway-Cook, A. (Eds.), *Development of posture and gait across the life span,* pp. 25-47. Columbia, SC: University of South Carolina Press.

Wiley, M.E., and Damiano, D.L. (1998) Lower-extremity strength profiles in spastic cerebral palsy. *Dev Med Child Neurol* 40: 100-107.

Woollacott, M.H., Burtner, P., Jensen, J., Jasiewicz, J., Roncesvalles, N., and Sveistrup, H. (1998) Development of postural responses during standing in healthy children and in children with spastic diplegia. *Neurosci Biobehav Rev* 22: 583-589.

Woollacott, M.H., and Shumway-Cook, A. (1986) The development of the postural and voluntary motor control system in Down's syndrome children. In Wade, M. (Ed.), *Motor skill acquisition of the mentally handicapped: Issues in research and training,* pp. 45-71. Amsterdam: Elsevier.

6

Impairment and Compensation of Reaching in Patients With Stroke and Cerebral Palsy

Mindy F. Levin

School of Rehabilitation, University of Montréal;
and Research Centre,
Rehabilitation Institute of Montréal

Carmen M. Cirstea, Philippe Archambault, and Florina Son

Neurological Science Research Centre,
University of Montréal;
and Research Centre,
Rehabilitation Institute of Montréal

Agnès Roby-Brami

Centre Nationale de la
Recherche Scientifique

The production of voluntary movement is a complex phenomenon that is not fully understood in either healthy or pathological systems. It is well-known that multijoint movements such as reaching are impaired after central nervous system (CNS) lesions involving motor areas and pathways. Although the basic capacity to produce movement is preserved in most patients (Bobath, 1978; Twitchell, 1951), impairments of motor control at single- and multijoint levels usually affect the quality of movement. This chapter summarizes recent research on motor control at the level of both single- and multijoint coordination in patients with adult-onset stroke and in children with cerebral palsy (CP).

The process of recovery from CNS lesions may be related to mechanisms permitting the system to use its inherent "redundancy." For example, when a joint is temporarily fixed at a specific angle by a splint or cast, compensatory movements in adjacent joints ensure that the task (i.e., reaching, gait) can still be accomplished. This is an example of redundancy in which the system is permitted to substitute other degrees of freedom (DOFs) for the ones usually recruited for the task. The concept of redundancy was introduced by Bernstein (1967) and refers to the capacity of the CNS to take advantage of the surplus number of muscles and joints of the musculoskeletal apparatus by selecting a desired trajectory and interjoint and intermuscle coordination from among many possible strategies to produce goal-directed movement. How the CNS uses this redundancy is fundamental to our understanding of motor control and motor recovery (Feldman and Levin, 1995; Kelso et al., 1993).

Related to redundancy is the phenomenon of plasticity in the CNS, for which considerable evidence exists, even in the adult CNS. Redundancy and plasticity are crucial to the process of recovery from CNS damage such as stroke or cerebral palsy. In stroke, although the most rapid recuperation may occur between 3 and 6 months postinjury (Lehman et al., 1975), recovery can proceed for longer periods up to 2 years (Skilbeck et al., 1983). Several mechanisms of CNS plasticity have been identified that may account for early and later recovery: the demarcation of latent intracortical synapses (Eccles, 1979), changes in synaptic efficiency (Tsukahara, 1981), and sprouting of intact neurons (Wall, 1980). In children, recovery is more difficult to quantify because it is integrated with maturational processes.

Motor recovery following damage to the primary motor cortex and/or the pyramidal tract may also be associated with functional reorganization in descending motor pathways unaffected by the lesion and bilaterally organized (Chollet et al., 1991; Fries et al., 1990, 1993; Hallett et al., 1998; Weiller et al., 1992). The basis for such reorganization may lie in the considerable functional overlap in cortical motor areas. For example, lesions of the anterior or posterior limb of the internal capsule, a common cause of hemiparesis, lead to initially severe motor impairment that is followed by different degrees of recovery with varying time courses (Danek et al., 1990; Fries et al., 1993). After initial severe hemiparesis, remarkable motor recovery can occur, even at the level of the control of finger movements (Fries et al., 1990). Some evidence suggests that recovery may be related to the appearance of ipsilateral motor pathways (but see Turton et al., 1996). Chollet et al. (1991) demonstrated that transcranial electrical stimulation of the damaged hemisphere evoked motor responses in both the contralateral and the ipsilateral hand. In addition, substantial coactivation of the ipsilateral motor cortex and contralateral nonprimary motor cortex was demonstrated by cortical metabolic activity levels in stroke patients moving their formerly paretic hand.

In the first part of this chapter, we will review evidence suggesting that, in some subjects with CNS lesions, the ability to take advantage of the system's redundancy may be limited at the local (single-joint) level due to deficits in the specification and regulation of stretch reflex thresholds in individual muscles (Jobin and Levin, 2000; Levin and Feldman, 1994) and in agonist/antagonist muscle pairs (Levin et al., 2000). This may provide an explanation of the mechanism underlying the range of movement limitations (Bobath, 1978; Brunnström, 1970) observed in stroke patients. In the second part of the chapter, we will review how the coordination of more complex movements involving several joints or body segments may be disrupted in pediatric and adult hemiparetic patients.

Deficits in the Regulation of Stretch Reflex Thresholds at the Level of Individual Muscles and Joints

Unilateral brain lesions resulting in hemiplegia are characterized by sensorimotor deficits such as spasticity, defined as hyperreflexia and hypertonia (Lance, 1980), and pathological synergies in the limbs contralateral to the hemispheric lesion (Bobath, 1978). In stroke, sensorimotor deficits include the inability to activate appropriate muscles (Bobath, 1978) and to coordinate movements between adjacent joints (Cirstea and Levin, 2000; Levin, 1996). Impairments such as weakness and loss of dexterity may be related to abnormal agonist motor unit activation (Colebatch et al., 1986; Tang and Rymer, 1981); altered mechanical properties of muscle fibers (Hufschmidt and Mauritz, 1985; Jakobsson et al., 1992); improper spatial and temporal muscle recruitment, including inappropriate agonist/antagonist coactivation (Gowland et al., 1992; Hammond et al., 1988; Levin and Dimov, 1997); and inability to control interactive torques generated by movements at more than one joint (Beer et al., 2000).

At the segmental level, motor abnormalities may be related to deficits in the organization of reflex activity such as reciprocal (Yanagisawa and Tanaka, 1978), presynaptic (Ashby and Verrier, 1976; Levin and Hui-Chan, 1992), and Renshaw cell inhibition (Veale et al., 1973). In addition to segmental mechanisms, alterations in signals descending from different cortical and subcortical brain regions have been implicated as mechanisms for the appearance of spasticity and the loss of coordinated movement in patients with brain injury (for reviews, see Burke, 1988; Wiesen-danger 1991). One such mechanism may be centrally driven alterations in the specification and regulation of stretch reflex thresholds at the level of individual muscles and in agonist/antagonist muscle pairs in specific joints. This has been demonstrated in the elbow joint in stroke subjects (Levin and Dimov, 1997; Levin and Feldman, 1994; Levin et al., 2000) and in children with spastic CP (Jobin and Levin, 2000).

The regulation of stretch reflex thresholds may be an essential result of supraspinal action and has important implications for spasticity and disordered motor control following CNS lesions. Matthews (1959) described a family of tension-length curves in the soleus muscle of the decerebrate cat elicited by a variety of segmental reflexes. Paralysis of gamma motoneurons resulted in changes in the tonic stretch reflex threshold without altering the shape of the tension-length relationship. Feldman and Orlovsky (1972) provided evidence that supraspinal structures have direct influences on tonic stretch reflex thresholds at segmental levels. They examined the effects of stimulation of various supraspinal structures (Dieter's nucleus, pyramidal tract, medial medullar reticular formation) in unanesthetized intercollicularly decerebrated cats. They found that continuous stimulation of Dieter's nucleus, which is excitatory to hindlimb extensors, resulted in a shift of the tension-length curve of the muscle to the left (i.e., initial tension developed at a shorter muscle length). This suggested a lowering of the stretch reflex threshold. Combining the excitatory input with that of different descending inhibitory systems resulted in further changes in the SR threshold without affecting the slope (reflex gain or stiffness) of the relationship. In decerebrated cats, pharmacologically elicited changes in the level of presynaptic inhibition of motoneurons also result in shifts of the SR threshold (Capaday, 1995). Changes in the slope of the force-length relationship were locked to the changes in the SR threshold, indicating that the SR gain was not regulated independently of the threshold.

In man, decreases in SR thresholds rather than increases in gains have been related to the appearance of spasticity (Lee et al., 1987; Powers et al., 1988). Spasticity has been defined as a motor disorder that refers, in part, to the clinical phenomenon of abnormally increased resistance to stretch of the passive muscle (Lance, 1980). In stroke patients, positional and velocity gains of the stretch reflex are reported to be increased (Thilmann et al., 1991) or unchanged (Lee et al., 1987; Powers et al., 1988) but are not correlated with the degree of clinically measured spasticity (Levin and Feldman, 1994; Levin et al., 2000). Thus, evidence suggests that the CNS may control movement by the regulation of stretch reflex thresholds. This is one of the essential underlying assumptions of the λ version of the equilibrium-point model of motor control (Feldman, 1966, 1986).

Specifically, the λ model suggests that the whole range in which muscle force and joint position are controlled is essentially defined by the range of regulation of stretch reflex thresholds. The model takes into account the velocity-dependence of the stretch reflex threshold and suggests that the coefficient of this dependency, the damping factor, may play an important role in stabilizing posture and movement. CNS lesions may disrupt the coordination of central commands and, consequently, interfere with the normal control of the threshold λ (Feldman, 1966, 1986; Feldman and Levin, 1993; Levin and Feldman, 1994). Disordered voluntary motor control, including spasticity and weakness, may result from consequences of altered central commands. These include (1) an increase in the background motoneuronal excitability leading to a shift in the lower limit of the threshold λ toward shorter muscle lengths (spasticity); (2) a decreased range of regulation of λ; (3) a change in the velocity sensitivity of λ; and (4) the inability of the system to overcome the background inhibition (disfacilitation) of motoneurons and thus decrease, when necessary, the SR threshold as required for muscle activation (weakness).

The first two disturbances have been measured in pediatric and adult hemiparetic patients in whom stretch reflex thresholds have been expressed in terms of joint angle instead of muscle length (see next paragraph). Clinical manifestations of these disturbances in hemiparetic patients include abnormal amounts of co-contraction in agonist and antagonist muscle groups (Bourbonnais et al., 1989), the fixed linear relationship between the amount of co-contraction and torque developed by paretic ankle dorsiflexors (Levin and Hui-Chan, 1994), and stereotypic movement patterns or synergies in the upper and lower limbs (Bobath, 1978). The third and fourth disturbances have also been measured in neurological populations (see next paragraph). The third may be clinically manifested as an increased velocity-sensitivity of the muscle to stretch, while the fourth may be associated with the appearance of weakness in specified joint ranges.

In healthy children and adults, stretch responses in relaxed elbow flexors and extensors are rarely, if ever, elicited (Levin and Feldman, 1994; Thilmann et al., 1991). This is interpreted in the λ model as indicating that the stretch reflex threshold (λ_+) is higher than the upper limit of the angular (anatomical) range of the joint (figure 6.1a). Voluntary movement, in the λ model, is produced by the regulation of the stretch reflex threshold from λ_- to λ_+, which extends throughout and beyond the angular range of the joint (θ_-, θ_+). The development of torque by an individual muscle is a consequence of the angular location of the threshold and the torque/angle relationship that is determined by descending and segmental neural pathways, as well as limb biomechanics. Our studies in hemiparetic adults and children suggest that the ability of subjects to relax muscles or to create torque and to balance an external load in different parts of the angular range is limited, possibly due to limitations in the range of regulation of stretch reflex thresholds in agonist and antagonist muscles acting around the joint (Jobin and Levin, 2000; Levin et al., 2000). For example, consider the case when λ_- lies inside of the angular range (θ_-, θ_+). In this case, the subject will be unable to generate nonzero force in the range of (θ_-, λ_-), resulting in muscle weakness in that range. Now consider the case in which $\lambda_+ < \theta_+$. In this range, the muscle would be unable to relax when stretched and spasticity would result. These pathologies can be observed in the same joint when the regulation of both limits of λ are deficient. Thus, weakness in one part of the anatomical range may be combined with spasticity in another part.

Figure 6.1b shows the velocity/angle phase diagrams and the static stretch reflex thresholds of the elbow flexors (λ^f) and elbow extensors (λ^e) in a stroke patient with moderate clinical symptoms. Phase diagrams and thresholds in a second patient with mild symptoms are shown in figure 6.1c. Static reflex thresholds cannot be directly measured since they would have to be measured when the velocity of stretch is equal to zero. Thus, we measured dynamic stretch reflex thresholds (λ^*) evoked by stretching the muscle at different velocities and then extrapolated these data to zero velocity to determine static thresholds. Each curved line

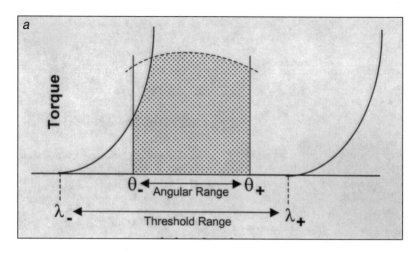

Joint Angle

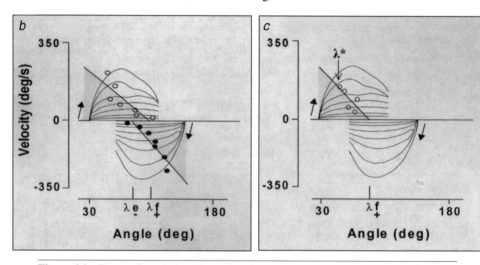

Figure 6.1 Range of regulation of stretch reflex thresholds. *(a)* According to the λ model, in individual muscles, the range of regulation of the stretch reflex threshold (λ_-, λ_+) extends beyond the angular range of the joint (θ_-, θ_+) so that position-dependent muscle torque can be regulated throughout the entire angular range (gray area). The position-dependent torque is determined by the torque/angle relationship of the muscle (thick curved lines) associated with each λ. This is illustrated theoretically for elbow flexors in *(a)*. *(b, c)* Velocity/angle phase diagrams for two adult hemiparetic subjects. The direction of the stretch is shown by arrows. Dynamic stretch reflex thresholds (λ*) for each mean velocity of stretch are illustrated for flexor (open circles) and extensor (closed circles) muscles. Static stretch reflex thresholds for flexors (λ_+^f) and extensors (λ_-^e) are determined by extrapolating the regression lines through the dynamic thresholds to zero velocity. Stretch reflex thresholds (λs) were found for flexors and extensors of the subject illustrated in *(b)* but only for the flexors of the subject in *(c)*. The area of active control is shown in gray for each example.

Reprinted from Levin and Feldman 1994.

on the phase diagrams represents the mean of eight randomly presented stretches. The stretch input consisted of nonconstant stretches having bell-shaped velocity profiles and mean velocities ranging from 160°/s to 8°/s. Electromyographic (EMG) activity in passive (noncontracting) elbow flexors and extensors, elbow velocity, and displacement were recorded during each stretch (Levin and Feldman, 1994; Levin et al., 2000). The angle and velocity corresponding to the threshold EMG activity in the stretched muscle represent the dynamic thresholds (λ^*) of that muscle for a given mean velocity of stretch (figure 6.1, b and c, circles). Stretch reflex thresholds in elbow flexors (open circles) measured in this way were found to lie within the angular range of the joint in both patients (i.e., $\theta_- < \lambda < \theta_+$). This was the case in almost all our children with CP ($n \sim 20$) and adult stroke patients ($n \sim 30$) tested to date. In nearly all children with CP and in about 33% of the adult stroke patients tested, the stretch reflex threshold was similarly limited in the elbow extensors (figure 6.1b).

The limitation in the specification and regulation of stretch reflex thresholds in agonist and antagonist muscles around the elbow may also lead to problems in joint stability. In a study in which preactivated elbow flexors were rapidly unloaded in stroke patients, 60% (6/10) had difficulty stabilizing the final position of the elbow (Levin and Dimov, 1997). In these patients, the transition from the initial preloaded state to the final state in which the arm was required to balance a new load was not accompanied by an appropriate level of tonic EMG activity in agonist and antagonist muscles (figure 6.2, a and b). This led to instability characterized by underdamped terminal oscillations about the final arm posture (figure 6.2d). This study suggested that those subjects with increased terminal oscillations were unable to specify a feedforward command leading to appropriate agonist/antagonist coactivation. We propose that the disturbance in specification of coactivation in appropriate ranges may be due to deficits in the coordinated regulation of stretch reflex thresholds of agonist and antagonist muscles. In the λ model, the control signal responsible for this coordination is termed the "C," or coactivation, command. The C command regulates agonist and antagonist thresholds appropriately in order to stabilize a joint (i.e., by increasing stiffness) or to provide a tonic level of activity for joint stability or precision during movement. The C command is likely issued by descending systems but mediated by segmental interneuronal circuitry (Feldman, 1993). Evidence for the supraspinal control of co-contraction has been demonstrated in the pre- and post-central cortex (Cheney et al., 1982; Humphrey and Reed, 1983), the cerebellum (Smith, 1981), and the red nucleus (Cheney and Mewes, 1986). Unlike the C command that likely does not influence the position of the equilibrium point (EP), at least in healthy subjects (Feldman, 1966, 1993), the other central command, R, coordinates changes of the agonist/antagonist thresholds in a different way. The R command produces movement by shifting the EP of the system and producing reciprocal agonist/antagonist muscle activation. The central origin of coactivation and reciprocal activation has been supported by findings of common and reciprocal drive to motor units of antagonist muscles during voluntary activation of thumb flexors and extensors in man (DeLuca and Mambrito, 1987).

In addition to deficits in the control of coactivation, limitations in the specification and range of regulation of reciprocal muscle activation have also been demonstrated in stroke patients (Levin et al., 2000). In a study in which patterns of elbow flexor and extensor muscle activation were identified during efforts to slowly flex or extend the elbow in different parts of the angular range, Levin et al. (2000) described limitations in the range of reciprocal innervation (range of R command

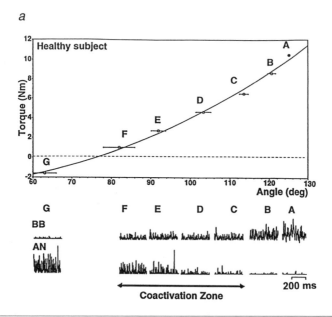

Figure 6.2 Measurement of the coactivation zone in the unloading experiments. *(a)* Example of an invariant characteristic (diagonal curved line) recorded in one healthy subject. Filled circle (A) shows the initial equilibrium point (EP). Final EPs following partial unloadings are shown as open circles (means ± SD). Tonic EMG levels (rectified signals) in agonist (biceps brachii, BB) and antagonist (anconeus, AN) muscles associated with each final EP (B to G) were measured during the last 400 ms of the holding period. EMG levels associated with the initial position were measured in a 200-ms period before unloading. Tonic agonist EMG activity decreased and antagonist EMG activity increased with the level of unloading. The angular zone in which agonist and antagonist muscles were simultaneously active is indicated by the double-headed arrow (Coactivation Zone). *(b)* Same as *(a)* for a stroke subject; however, in this subject, tonic agonist activity did not decrease in the holding period, and antagonist EMG was absent. *(c)* Averaged (*n* = 10) kinematic responses to five different levels of unloading in a healthy subject similar to the one shown in *(a)*. The initial torque was 8 Nm. Responses changed systematically with the level of unloading. The peak velocity, time to peak velocity, and final limb position were monotonic functions of the size of the unloading step. Especially obvious from the phase diagram (not shown) is that the magnitude of the dynamic overshoot increased with the unloading step. *(d)* Same as *(c)* for the stroke subject. As in the healthy subject, kinematic responses changed systematically with the level of unloading, but they were characterized by oscillations that increased with the level of unloading.
Reprinted from Levin and Dimov 1997.

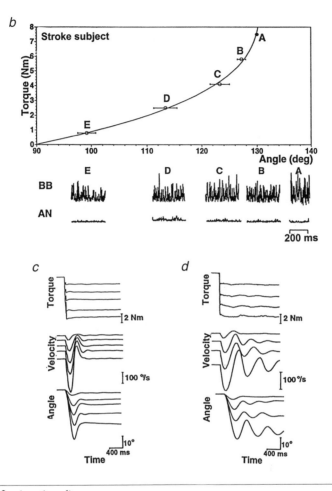

Figure 6.2 *(continued)*

according to the λ model) in all but 1 out of 12 stroke patients. In the healthy subjects (figure 6.3a), both the agonist and antagonist λs could be regulated throughout and beyond the angular limit of the joint. This was evidenced by the presence of a reciprocal pattern of agonist/antagonist muscle innervation during free low-velocity voluntary movements throughout the joint range in each direction (elbow flexion/extension). Therefore, the range of regulation of R (R_-, R_+) was assumed to be greater than the range defined by the anatomical limits of the joint (θ_-, θ_+). Indeed, the measurable range of the R command, shown in figure 6.3, is restricted by the mechanical limits of the joint. In stroke patients, the range of R was decreased and the border(s) fell within the anatomical limits of the joint. The range of R was defined as the range in which movement occurred by activating either the flexors or the extensors while the antagonist muscle remained silent. The borders of the range

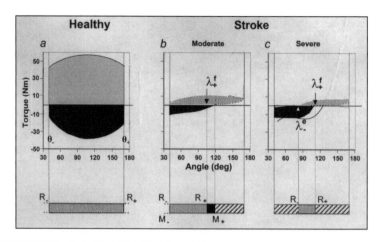

Figure 6.3   The areas in which active torque was generated in the flexor (gray shading) and extensor (black shading) muscles in *(a)* one healthy subject and *(b, c)* two stroke subjects. Second-order polynomial curves were fit to the data representing the maximal torques that could be generated at each joint angle tested. The inferior limits of each area are marked by lines indicating passive torque (measured when each muscle group was extended at a slow velocity, < 8°/s). Filled and open arrows represent static stretch reflex thresholds for the flexors and the extensors, respectively. *(a)* In the healthy subject, substantial torque could be generated at each joint angle for both muscle groups, and there was no appreciable contribution of passive torque. *(b)* For one stroke subject, flexor torque could be produced in the whole angular range, but extensor torque could be produced in only part of the range. Passive flexor torque occurred when the elbow was brought into extension. In *(c)*, the range of active torque production was limited in both directions. Passive flexor and extensor torques were present at the extremes of each range. Horizontal bars below each panel represent the range of reciprocal innervation, R (gray bars), the range of active movement (M_, M_+), and the ranges in which no movement or movement in only one direction was possible (hatched bars). In the healthy subject and stroke subject shown in *(c)*, the range of M corresponds to the range of R, but in the stroke subject shown in *(b)*, the range extends beyond the range of R (black bar). The inferior border of R (R_) extends beyond the lower limit of the angular range (θ_, ≈ 30°) *(a)* in the healthy subject and *(b)* in one stroke subject. In the other stroke subject *(c)*, the range of R is limited, and movement in only one direction is possible within the hatched ranges.

of R were defined as the static thresholds of the flexors (λ^f_+) and the extensors (λ^e_-, figure 6.3c) or the flexors and the lower limit of the angular range of the joint (θ_, figure 6.3b). Although movement of the limb beyond the borders of the range of R was possible, it was accompanied by coactivation rather than a reciprocal pattern of muscle innervation. The total range of active motion, often extending beyond the range of R, was defined as the range of M (figure 6.3b; black section of horizontal bar). The coactivation occurred because the active movement brought the arm beyond the limit of regulation of λ for the antagonist so that the antagonist muscle was reflexly activated.

Our findings of abnormal patterns of reciprocal activation and coactivation, as well as the presence of spasticity and the lack of stability in the hemiparetic arm, are consistent with numerous reports in the literature (see introductory paragraphs). The λ model provides a means of explaining these deficits based on the mechanism of the specification and regulation of stretch reflex thresholds in individual muscles and joints. However, the motor disorder in stroke extends beyond problems at the individual joint level. Indeed, deficits in interjoint coordination (Levin, 1996; Trombly, 1992), timing and sequencing of the recruitment of multiple body segments within a movement (Archambault et al., 1999; Levin et al., 2002; Roby-Brami et al., 1997), and anticipatory postural adjustments (Dean and Shepherd, 1997; Michaelsen et al., 2001) are also characteristic of the motor disorder in stroke and in other CNS pathologies.

Abnormal Coordination of Movement and Compensations

Dewald et al. (1995) have suggested that abnormal directional specificity of activation patterns of individual muscles in the hemiparetic arm may account for the presence of abnormal movement synergies. Abnormal flexor or extensor synergies, originally described by Twitchell (1951), are associated with characteristic movement sequences when the patient attempts to move the affected limb (see Brunnström, 1970, for a description). Such abnormal synergies are often observed in patients in the early stages after stroke, and their influence may diminish during the process of recovery. Presumably, these can be characterized as "positive features" associated with exaggerated supraspinal drive following stroke (Burke, 1988; Wiesendanger, 1991). Deficits in the regulation of λs so that muscles of adjacent joints are constrained to act together in certain ranges may also explain the appearance of abnormal movements and movement synergies. This theoretical construct is illustrated in figure 6.4. For simplicity, a theoretical reaching movement composed of only three DOFs is illustrated: wrist extension (E_w), elbow extension (E_e), and shoulder horizontal adduction (Add). The task requires that the subjects reach in a horizontal plane from an initial position in which the hand is in front of the sternum and the shoulder is flexed and abducted (EP_1) to a final position (EP_2) in the sagittal direction. The ranges of R and M for all three joints are illustrated by horizontal bars. Let us assume that the executive command involves the coordination of the three joints at some suprasegmental level so that the deficits in the ranges of R (gray areas) and M (black areas) are as illustrated in figure 6.4. The executive command (dashed lines) specifies the ranges of movement at each joint (thick arrows). The required movement is possible at the wrist joint, and the resultant movement (thin arrow) matches the command. However, due to limitations in the ranges of R, movements are not matched at the elbow and shoulder joints, and the resultant arm reach is made with limited elbow extension and shoulder horizontal adduction.

The limitations in the ranges of active movement due to decreased ranges of R result in the movement not being made as intended. Our studies of stroke patients

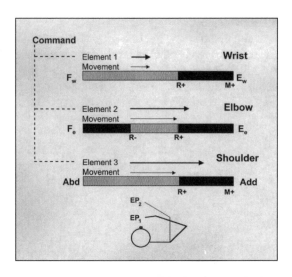

Figure 6.4 Theoretical diagram explaining the limitation of movement in a stroke patient making a hypothetical reaching movement from EP_1 to EP_2. Horizontal bars represent the range of active motion for each joint bounded on one end by extreme flexion (F_w, F_e) or abduction (Abd) and on the other by extreme extension (E_w, E_e) or adduction (Add). For simplicity, only one degree of freedom for each joint is considered. The ranges of R (gray areas) and M (black areas) are the same as in figure 6.3 but are shown here for the wrist, the elbow, and the shoulder. The thick arrows indicate the range of movement specified by the executive command (dashed lines) for each joint. The thin arrows indicate the movements resulting from the command, taking into account the limitations in R and M for each joint. The diagram shows that if the required movement extends into a range beyond the range of R in a joint, the resultant range of movement (e.g., the elbow) will be decreased.

and children with CP show that reaching movements may still be accomplished despite these movement limitations by use of a compensatory strategy (the recruitment of the trunk), even when the target is located within the boundary of arm's length (Cirstea and Levin, 2000; Cirstea et al., 1998; Michaelsen et al., 2001; Roby-Brami et al., 1997). This is observed even though movement of the trunk is not necessary to accomplish the task. Examples of arm movements made by two adult stroke patients and two children with CP are shown in figure 6.5. The movement consisted of an unconstrained three-dimensional reaching task from an initial position located approximately 15 cm [5.9 in.] from the midline of the sternum to a final position involving grasping a cone (adults) or touching a toy (children) located well within the reach of the arm (three-quarters arm's length). Movement trajectories are contrasted with those from healthy subjects in each row.

In general, movements in patients were characterized by spatial and temporal segmentation, suggesting the loss of a smooth coordination between adjacent joints. Although no patients had limitations in the passive joint ranges (no joint contractures) of either the elbow or the shoulder, the more severely affected patients used smaller

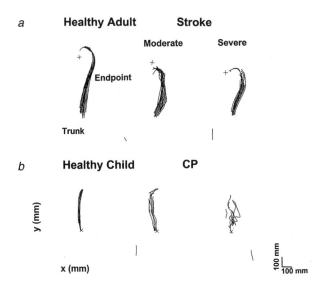

Figure 6.5 Horizontal projection of three-dimensional recordings of hand (endpoint) and trunk trajectories *(a)* for grasping in adults and *(b)* pointing in children when the task requires reaching to a point within arm's length. The task was similar to the one depicted in figure 6.4. The results in healthy control subjects are compared to those in patients with moderate and severe impairments. The hand and trunk trajectories are reconstructed based on data from markers attached to the distal end of the second metacarpal and the sternum, respectively. Each panel includes 10-20 individual trajectories for the hand (endpoint) and one trajectory for the trunk. The x direction is frontal and the y direction is sagittal.

active ranges of motion in each of these joints to accomplish the reaching task. This is illustrated for stroke patients in figure 6.6. The decreased range of joint motion may be explained by several possible mechanisms: weakness of the agonist muscles (limitation in the range of R), excessive spastic restraint from the antagonist muscle (i.e., movement into a range extending beyond the threshold of λ for the antagonist muscle), or instability of the joint in the extended position (inappropriate specification of the C command).

However, in spite of the limitations in active joint range, even the most clinically disabled patients moved the endpoint (the hand) to the target position. This was accomplished by the recruitment of an additional DOF, the trunk, not normally involved in the task of reaching objects placed close to the body (figure 6.5, thin single lines below each group of arm trajectories). The appearance of such a compensatory mechanism underscores the ability of the patients to make use of the remaining redundancy in the damaged CNS.

Arm-pointing movements involving the trunk in healthy subjects have been investigated in a number of studies (Adamovich et al., 1998; Kaminski et al., 1995; Ma and Feldman, 1995; Pigeon and Feldman, 1998; Saling et al., 1996). During natural trunk-assisted pointing and grasping movements to targets beyond arm's

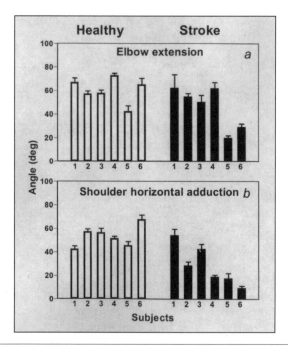

Figure 6.6   *(a)* Maximal elbow extension and *(b)* shoulder horizontal adduction observed during reaching movement directed to an object located at three-quarters arm's length. Each column represents the mean + SD of 10 trials for healthy subjects (white columns) or stroke patients (black columns).

reach, the trunk begins to move at the same time or before the hand and stops moving after the end of hand movement (Kaminski et al., 1995; Saling et al., 1996). The hand stops and is maintained in the same position while the trunk continues to move. This indicates that the effects of trunk motion on the hand position were adequately neutralized by compensatory joint rotations at the elbow and shoulder. Otherwise, a clear effect of the trunk movement on the hand trajectory would have been observed, and it would have been impossible for the hand to stop moving before the trunk. This compensation appears to be preserved in stroke patients and in children with CP. Figure 6.7 shows the timing of hand and trunk movements at the beginning and end of reaching. For healthy subjects, the task involves reaching to a target placed beyond the length of the arm so that trunk bending is required. For subjects with stroke (affected arm) and CP (both arms), data are shown from reaches to targets placed at approximately three-fourths the length of the arm for which subjects abnormally recruited trunk movement. Figure 6.7 illustrates that despite prolongations in total movement time, hand-trunk timing during reaching tasks normally or abnormally involving the trunk was preserved in these patients. This implies that the command signal specifying the timing of the recruitment of body segments may not be affected by the CNS lesion. This may not be surprising since timing and

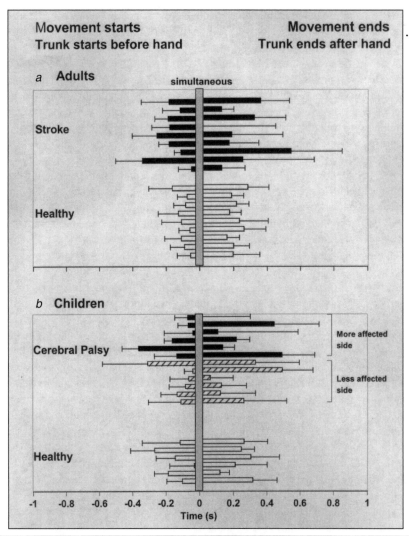

Figure 6.7 Hand-trunk timing at the start and end of the reaching movement. Each horizontal bar corresponds to one subject. Bars projecting to the left indicate the delay (–SD) in the initiation of movement between the trunk and the hand at movement start. Those projecting to the right show the delay (+SD) between the termination of hand and trunk displacement at movement end. The middle gray area corresponds to events occurring simultaneously (±20 ms). Healthy subjects and patients had similar hand-trunk timing despite the difference in the range of trunk movement (see figure 6.5) and the speed of movement. *(a)* Hand-trunk timing observed in adult stroke patients reaching with their affected arms and in healthy subjects reaching with their dominant arms to a target placed within arm's length. *(b)* Hand-trunk timing observed in CP children reaching with their more affected and less affected arms and in healthy children reaching with their dominant arms to an object placed within arm's length.

sequencing of movements are likely controlled at subcortical as well as cortical brain levels (Kalaska et al., 1990; Winstein et al., 1997; Wise and Strick, 1984), and subcortical areas may be relatively intact in our patients.

When healthy subjects deliberately recruit the trunk during a reaching task not normally involving trunk motion, the trunk motion does not affect the hand trajectory (Ma and Feldman, 1995). These authors hypothesize that this particular task involves two synergies: one moving the arm joints displacing the hand to the target (reaching synergy) and the other moving the trunk and arm joints without affecting the position of the hand (compensatory synergy). This process is independent of visual feedback (Pigeon and Feldman, 1998). In theory, such a compensatory synergy could either be produced by the central commands or mediated through afferent feedback. To distinguish between these two possibilities, Adamovich et al. (1998) analyzed fast hand-trunk movements without vision in which trunk movement was sometimes prevented. Only minor changes were found in the hand trajectory, hand velocity profile, movement precision, and variability in response to the perturbation. Thus, blocking the trunk also stopped the compensatory synergy; otherwise, a deflection of the endpoint trajectory from that in the unperturbed condition would have been observed. This implies that the compensatory hand-trunk coordination is guided mainly by afferent rather than central signals. The authors concluded that afferent signals originating from the vestibular apparatus due to head movement or from proprioceptive signals of the moving trunk or shoulder may play a role in the integration of multiple body segments into the motor task.

Whether or not patients with neurological disorders are able to use afferent signals in the same way as healthy subjects to produce coordinated multisegmental movement is unknown and merits further study. As we have shown, stroke patients and children with CP substitute trunk motion for decreased active elbow extension and shoulder adduction. Hand and trunk recruitment, however, occurs in the same sequence as in healthy subjects reaching to targets placed beyond arm's reach. The difference is that the trunk is recruited in hemiparetic patients and children with CP for targets placed within arm's reach. This may indicate that the additional trunk movement is a natural response to the limitations in the extension of the arm. The CNS may integrate the limitations in arm movement by lowering the threshold of trunk recruitment, thus making use of the additional DOFs in the body to achieve its goal (Levin et al., 2002). Interestingly, when the trunk motion was restrained by the use of a harness in stroke patients, they were able to use a greater range of elbow extension to transport the hand to the target (Michaelsen et al., 2001; see figure 6.8). This suggests that the mechanism triggering compensatory adaptation may be preserved in stroke patients.

Conclusion

Stroke patients have deficits in the ability to use certain ranges of elbow joint motion when reaching into external space. These deficits may be explained by limitations in the range of regulation of stretch reflex thresholds measured in elbow flexor and

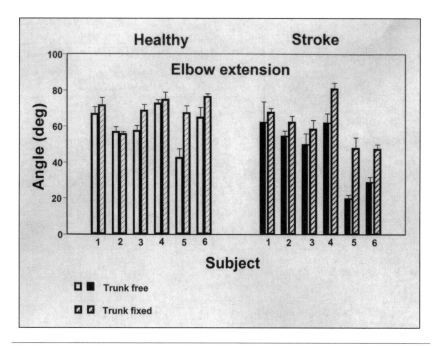

Figure 6.8 Maximal elbow extension observed during a reaching and grasping movement to an object located within arm's length with the trunk free (solid bars) or fixed (hatched bars). Each column represents the mean +SD of 10 trials for each subject.

extensor muscles in hemiparetic adults and children. Further deficits may reside in the specification of command signals producing reciprocal motion and/or coactivation about a joint. Control deficits at several individual joints may be combined when patients attempt to produce voluntary movement, and abnormal coordination may occur. For reaching movements into external space, these may be offset by the appearance of compensatory movement involving the trunk. We suggest that hemiparetic patients are able to use compensatory synergies that are part of the normal repertoire of possible movement in healthy subjects. Our data show that trunk-assisted reaching occurs earlier in stroke patients but that patients may retain the capacity to use this mechanism to compensate arm weakness or instability. These findings have important implications for rehabilitation since they suggest that the damaged CNS may retain a greater capacity for redundancy and compensation than has been generally assumed.

Acknowledgments

The authors wish to thank students Marius Dimov, Anamaria Luta, Ruud Selles, and Martine Verhuel, as well as Serge Adamovich, Onno Meijer, and Anatol

Feldman, who have all contributed to this work. The studies summarized here have been supported by the Foundation of the Rehabilitation Institute of Montréal, the Natural Science and Engineering Research Council of Canada, the Medical Research Council of Canada, and Institut National de la Santé et de la Recherche Médicale-MRC.

References

Adamovich, S.V., Levin, M.F., Archambault, P., Poizner, H., and Feldman, A.G. (1998) Compensation in arm-trunk coordination for perturbation during reaching. *Soc Neurosci Abstr* 24: 2109.

Archambault, P., Pigeon, P., Feldman, A.G., and Levin, M.F. (1999) Recruitment and sequencing of different degrees of freedom during pointing movements involving the trunk in healthy and hemiparetic subjects. *Exp Brain Res* 126: 55-67.

Ashby, P., and Verrier, M. (1976) Neurophysiologic changes in hemiplegia. *Neurology* 26: 1145-1151.

Beer, R.F., Dewald, J.P.A., and Rymer, W.Z. (2000) Deficits in the coordination of multijoint arm movements in patients with hemiparesis: Evidence for disturbed control of limb dynamics. *Exp Brain Res* 131: 305-319.

Bernstein, N.A. (1967) *The Co-ordination and regulation of movements*. Oxford: Pergamon Press.

Bobath, B. (1978) *Adult hemiplegia. Evaluation and treatment*, 2nd ed. London: Heinemann Medical.

Bourbonnais, D., Vanden-Noven, S., Carey, K.M., and Rymer, W.Z. (1989) Abnormal patterns of elbow muscle activation in hemiparetic subjects. *Brain* 112: 85-102.

Brunnström, S. (1970) *Movement therapy in hemiplegia. A neurophysiological approach*. New York: Harper & Row.

Burke, D. (1988) Spasticity as an adaptation to pyramidal tract injury. In Waxman, S.G. (Ed.), *Advances in neurology. Functional recovery in neurological disease, vol. 47*, pp. 401-423. New York: Raven Press.

Capaday, C. (1995) The effects of baclofen on the stretch reflex parameters of the cat. *Exp Brain Res* 104: 287-296.

Cheney, P.D., Kasser, R., and Holsapple, J. (1982) Reciprocal effect of single corticomotoneuronal cells on wrist extensor and flexor muscle activity in primate. *Brain Res* 247: 164-168.

Cheney, P.D., and Mewes, K.H. (1986) Properties of rubromotoneuronal cells studied in the awake monkey. *Abstracts of Symposium on Neural Control of Limb Movement*, Seattle, WA, A-16.

Chollet, F., DiPiero, V., Wise, R.J.S., Brooks, D.J., Dolan, R.J., and Frankowiak, R.S.J. (1991) The functional anatomy of motor recovery after stroke in humans: A study with positron emission tomography. *Ann Neurol* 29: 63-71.

Cirstea, M.C., Leduc, B., and Levin, M. (1998) Hemiparetic patients recruit additional degrees of freedom to compensate lost motor function. In Arsenault, B., McKinley, P., and McFadyen, B. (Eds.), *Proc. of the Twelfth Congress of the International Society of Electrophysiology and Kinesiology* (pp. 106-107). Montréal: ISEK.

Cirstea, M.C., and Levin, M.F. (2000) Compensatory strategies for reaching in stroke. *Brain* 123: 940-953.

Colebatch, J.G., Gandevia, S.C., and Spira, P.J. (1986) Voluntary muscle strength in hemiparesis: Distribution of weakness at the elbow. *J Neurol Neurosurg Psychiatry* 49: 1019-1024.

Danek, A., Bauer, M., and Fries, W. (1990) Tracing of neuronal connections in the human brain by magnetic resonance imaging in vivo. *Eur J Neurosci* 2: 112-115.

Dean, C.M., and Shepherd, R.L. (1997) Task-related training improves performance of seated reaching tasks after stroke. A randomized controlled trial. *Stroke* 28: 722-728.

DeLuca, C.J., and Mambrito, B. (1987) Voluntary control of motor units in human antagonist muscles: Coactivation and reciprocal activation. *J Neurophysiol* 58: 525-542.

Dewald, J.P.A., Pope, P.S., Given, J.D., Buchanan, T.S., and Rymer, W.Z. (1995) Abnormal muscle coactivation patterns during isometric torque generation at the elbow and shoulder in hemiparetic subjects. *Brain* 118: 495-510.

Eccles, J.C. (1979) Synaptic plasticity. *Naturwissenschaften* 66: 147-153.

Feldman, A.G. (1966) Functional tuning of the nervous system with control of movement and maintenance of a steady posture: II. Controllable parameters of the muscle. *Biophysics* 11: 565-578.

Feldman, A.G. (1986) Once more on the equilibrium-point hypothesis (λ model) for motor control. *J Mot Behav* 18: 17-54.

Feldman, A.G. (1993) The coactivation command for antagonist muscles involving Ib interneurons in mammalian motor control systems. An electrophysiologically testable model. *Neurosci Lett* 155: 167-170.

Feldman, A.G., and Levin, M.F. (1993) Control variables and related concepts in motor control. *Concepts Neurosci* 4: 25-51.

Feldman, A.G., and Levin, M.F. (1995) The origin and use of positional frames of reference in motor
control. *Behav Brain Sci* 18: 723-744.

Feldman, A.G., and Orlovsky, G.N. (1972) The influence of different descending systems on the tonic
stretch reflex in the cat. *Exp Neurol* 37: 481-494.

Fries, W., Danek, A., Bauer, W.M., Witt, T.N., and Leinsinger, G. (1990) Hemiplegia after lacunar stroke
with pyramidal degeneration shown in vivo: A model for functional recovery. In Wild, K. von,
and Janzik, H. (Eds.), *Neurologische Fruhrehabilitation,* pp. 11-17. Munich: Zuchschwerdt.

Fries, W., Danek, A., Scheidtmann, K., and Hamburger, C. (1993) Motor recovery following capsular
stroke. Role of descending pathways from multiple motor areas. *Brain* 116: 369-382.

Gowland, C., deBruin, H., Basmajian, J.V., Plews, N., and Burcea, I. (1992) Agonist and antagonist
activity during voluntary upper-limb movement in patients with stroke. *Phys Ther* 72: 624-633.

Hallett, M., Wassermann, E.M., Cohen, L.G., Chmielowska, J., and Gerloff, C. (1998) Cortical mechanisms of recovery of function after stroke. *Neuro Rehab* 10: 131-142.

Hammond, M.C., Fitts, S.S., Kraft, G.H., Nutter, P.B., Trotter, M.J., and Robinson, L.M. (1988) Co-contraction
in the hemiparetic forearm: Quantitative EMG evaluation. *Arch Phys Med Rehab* 69: 348-351.

Hufschmidt, A., and Mauritz, K.-H. (1985) Chronic transformation of muscle in spasticity: A peripheral
contribution to increased tone. *J Neurol Neurosurg Psychiatry* 48: 676-685.

Humphrey, D.R., and Reed, D.J. (1983) Separate control systems for control of joint movement and joint
stiffness: Reciprocal activation and coactivation of antagonist muscles. In Desmedt, J.E. (Ed.),
Motor control mechanisms in health and disease, pp. 347-373. New York: Raven Press.

Jakobsson, F.L., Grimby, L., and Edstrom, L. (1992) Motoneuron activity and muscle fibre type com-
position in hemiparesis. *Scand J Rehab Med* 24: 115-119.

Jobin, A., and Levin, M.F. (2000) Regulation of stretch reflex threshold in elbow flexors in children with
cerebral palsy: A new measure of spasticity. *Dev Med Child Neurol,* 42: 531-540.

Kalaska, J.F., Cohen, D.A.D., Prud'homme, M., and Hyde, M.L. (1990) Parietal area 5 neuronal activity
encodes movement kinematics, not movement dynamics. *Exp Brain Res* 80: 351-364.

Kaminski, T.R., Bock, C., and Gentile, A.M. (1995) The coordination between trunk and arm motion
during pointing movements. *Exp Brain Res* 106: 457-466.

Kelso, J.A.S., Buchanan, J.J., DeGuzman, G.C., and Ding, M. (1993) Spontaneous recruitment and
annihilation of degrees of freedom in biological coordination. *Phys Lett A* 179: 364-371.

Lance, J.W. (1980) The control of muscle tone, reflexes, and movement. Robert Wartenberg Lecture.
Neurol 30: 1303-1313.

Lee, W.A., Boughton, A., and Rymer, W.Z. (1987) Absence of stretch reflex gain enhancement in
voluntarily activated spastic muscle. *Exp Neurol* 98: 317-335.

Lehman, J.F., Delauter, B.J., and Fowler, R.S. (1975) Stroke: Does rehabilitation affect outcome? *Arch Phys Med Rehabil* 56: 375-382.

Levin, M.F. (1996) Interjoint coordination during pointing movements is disrupted in spastic hemiparesis. *Brain* 119: 281-294.

Levin, M.F., and Dimov, M. (1997) Spatial zones for muscle coactivation and the control of postural
stability. *Brain Res* 757: 43-59.

Levin, M.F., and Feldman, A.G. (1994) The role of stretch reflex threshold regulation in normal and
impaired motor control. *Brain Res* 637: 23-30.

Levin, M.F., and Hui-Chan, C.W.Y. (1992) Relief of hemiparetic spasticity by TENS is associated with
improvements in reflex and voluntary functions. *Electroenceph Clin Neurophysiol* 82: 131-142.

Levin, M.F., and Hui-Chan, C.W.Y. (1994) Ankle spasticity is inversely correlated with antagonist
voluntary contraction in hemiparetic subjects. *Electromyogr Clin Neurophysiol* 34: 415-425.

Levin, M.F., Michaelsen, S., Cirstea, C., and Roby-Brami, A. (2002) Trunk involvement during reaching
in adult hemiparesis. *Exp Brain Res* (in press).

Levin, M.F., Selles, R.W., Verheul, M.H.G., and Meijer, O.G. (2000) Deficits in the coordination of agonist and antagonist muscles in stroke patients: Implications for normal motor control. *Brain Res* 853: 352-369.

Ma, S., and Feldman, A.G. (1995) Two functionally different synergies during arm reaching movements involving the trunk. *J Neurophysiol* 73: 2120-2122.

Matthews, P.B.C. (1959) A study of certain factors influencing the stretch reflex of the decerebrate cat. *J Physiol (Lond)* 147: 547-564.

Michaelsen, S., Luta, A., Roby-Brami, A., and Levin, M.F. (2001) Effect of trunk restraint on the recovery of reaching movements in hemiparetic patients. *Stroke* 32: 1875-1883.

Pigeon, P., and Feldman, A.G. (1998) Compensatory arm-trunk coordination in pointing movements is preserved in the absence of visual feedback. *Brain Res* 802: 274-280.

Powers, R.K., Marder-Meyer, J., and Rymer, W.Z. (1988) Quantitative relations between hypertonia and stretch reflex threshold in spastic hemiparesis. *Ann Neurol* 23: 115-124.

Roby-Brami, A., Fuchs, S., Mokhtari, M., and Bussel, B. (1997) Reaching and grasping strategies in hemiparetic patients. *Motor Control* 1: 72-91.

Saling, M., Stelmach, G.E., Mescheriakov, S., and Berger, M. (1996) Prehension with trunk assisted reaching. *Behav Brain Res* 80: 753-760.

Skilbeck, C.E., Wade, D.T., Hewer, K.L., and Wood, V.A. (1983) Recovery after stroke. *J Neurol Neurosurg Psychiatry* 46: 5-8.

Smith, A.M. (1981) The coactivation of antagonist muscles. *Can J Physiol Pharmacol* 59: 733-747.

Tang, A., and Rymer, W.Z. (1981) Abnormal force-EMG relations in paretic limbs of hemiparetic human subjects. *J Neurol Neurosurg Psychiatry* 44: 690-698.

Thilmann, A.F., Fellows, S.J., and Garms, E. (1991) The mechanism of spastic hypertonus. Variation in reflex gain over the time course of spasticity. *Brain* 114: 233-244.

Trombly, C.A. (1992) Deficits of reaching in subjects with left hemiparesis: A pilot study. *Am J Occup Ther* 46: 887-897.

Tsukahara, N. (1981) Synaptic plasticity in the mammalian central nervous system. *Ann Rev Neurosci* 4: 351-379.

Turton, A., Wroe, S., Trepte, N., Fraser, C., and Lemon, R.N. (1996) Contralateral and ipsilateral EMG responses to transcranial magnetic stimulation during recovery of arm and hand function after stroke. *Electroenceph Clin Neurophysiol* 101: 316-328.

Twitchell, T.E. (1951) The restoration of motor function following hemiplegia in man. *Brain* 74: 443-480.

Veale, J.L., Rees, S., and Mark, R.F. (1973) Renshaw cell activity in normal and spastic man. In Desmedt, J.E. (Ed.), *New developments in electromyography and clinical neurophysiology, vol. 3*, pp. 523-537. Basel: Karger.

Wall, P.D. (1980) Mechanisms of plasticity of connections following damage in adult mammalian nervous system. In Bach-y-Rita, P. (Ed.), *Recovery of function: Theoretical considerations for brain injury rehabilitation*, pp. 91-105. Baltimore: Park Press.

Weiller, C., Chollet, F., Friston, K.J., Wise, R.J.S., and Franckowiak, R.S.J. (1992) Functional reorganization of the brain in recovery from striatocapsular infarction in man. *Ann Neurol* 31: 463-472.

Wiesendanger, M. (1991) Weakness and the upper motoneurone syndrome: A critical pathophysiological appraisal. In Berardelli, A., Benecke, R., Manfredi, M., and Marsden, C.D. (Eds.), *Motor disturbances II*, pp. 319-332. New York: Academic Press.

Wise, S.P., and Strick, P.L. (1984) Anatomical and physiological organization of the non-primary motor cortex. *TINS* 7: 442-446.

Winstein, C.J., Grafton, S.T., and Pohl, P.S. (1997) Motor task difficulty and brain activity—investigation of goal-directed reciprocal aiming using positron emission tomography. *J Neurophysiol* 77: 1581-1594.

Yanagisawa, N., and Tanaka, R. (1978) Reciprocal Ia inhibition in spastic paralysis in man. In Cobb, W.A., and van Duijn, H. (Eds.), *Contemporary clinical neurophysiology*, EEG Suppl. 34, pp. 521-526. Amsterdam: Elsevier.

Reorganization of Motor Patterns During Motor Learning

A Specific Role of the Motor Cortex

M. Ioffe
Institute of Higher Nervous Activity and Neurophysiology, Russian Academy of Sciences

J. Massion and C. Schmitz
UPR Neurobiologie et Mouvement, Centre National de la Recherche Scientifique, Marseille, France

F. Viallet and R. Gantcheva
Service de Neurologie, CHG d'Aix-eu-Provence, France

A role of the motor cortex (MCx) in motor learning is being broadly discussed in the current literature. A particular function of MCx in motor learning is suppression of synergies and coordination that interfere with acquisition of new motor patterns.

Experimental animal models based on inhibition of certain natural synergies or reflexes in the process of learning new coordination have been developed in which the MCx is responsible for the inhibition of natural motor patterns. After MCx lesions, the natural synergies dominate again, and the learned movement cannot be adequately performed. No recovery could be observed even after several years of retraining. Similar disturbances occur after combined lesions of the premotor and parietal associative cortices or after lesions of the cerebellar nuclei. However, after the associative cortex or cerebellar lesions, the recovery of learned coordination is possible during several weeks of retraining. Thus, inhibition of inappropriate synergies or coordination during motor learning is a specific function of MCx. This was shown not only for limb movements but also for postural adjustments. MCx participates in organizing new coordination between posture and movement in humans as well. An acquisition of a posturo-kinetic coordination in a bimanual load-lifting task has been dramatically disturbed in patients with capsular hemiparesis. The defect is less obvious in Parkinson's patients.

Motor Learning: Operating Characteristics

Although mechanisms of motor learning have been studied extensively, motor learning still lacks a solid universal definition. At present, three forms of motor learning are usually distinguished: adaptation learning, skill learning, and conditional learning (Donoghue et al., 1996; Hallett et al., 1996). Assuming that motor learning involves development or elaboration of new movements, only skill learning should perhaps be considered as proper motor learning. However, even in the case of skill learning, it is sometimes difficult to determine whether it involves learning of a new movement or modification of a previously learned one. For example, according to Bernstein (1967), skill learning represents a reduction of number of degrees of movement freedom. However, this general principle includes skill improvement by practice as well and is not always clearly observed in particular cases of skill learning. Meanwhile, there is some evidence that learning a new skill and performing a previously learned one involve different sets of activated brain structures (van Mier et al., 1998).

To differentiate between new and modified skills at the functional level, a principle of operating characteristics has been used. "An operating characteristic describes a set of movements that relate different movement variables to one another" (Hallett et al., 1996). According to this approach, a feature of skill learning is a change of operating characteristic during a motor performance. A well-known example is Fitts's law (Donoghue et al., 1996; Fitts, 1954; Hallett et al., 1996). According to Fitts's law, the greater the velocity of a movement, the less accurate it is. Thus, when Fitts's law is applied to repeated trials, learning a new skill is not involved. Only if both parameters are nonreciprocally changed should skill learning be considered (Donoghue et al., 1996).

Other operating characteristics may play a role in motor learning as well, in particular, the reduction of number of degrees of freedom, according to Bernstein, or a coordination of two different movements during learning (Jaric and Latash, 1998). Another operating characteristic is based on reorganization of coordination in the process of motor learning (Ioffe, 1973; Ioffe et al., 1988).

During motor learning, some innate or well-learned synergies may be incorporated into the pattern of a new movement. However, in some cases, natural synergies may interfere with a movement being learned, and thus they have to be inhibited during its performance. The inhibition of inappropriate synergies or coordinations has to be learned as well. An example of such a situation is a simultaneous rotation of both hands or forearms in the sagittal plane, which may be easily performed in the same direction but requires training for other directions. A natural coordination providing rotation in the same direction opposes the coordination required for rotation in different directions and has to be suppressed in the process of learning.

Thus, a reorganization of natural coordinations, and particularly inhibition of inappropriate motor patterns interfering with a movement being learned, is a very important operating characteristic for elaboration of new coordinations during motor learning.

Reorganization of Natural Coordinations During Motor Learning: Animal Models

Reorganization of innate motor reactions during instrumental learning in animals has been studied in the pioneer work of Zelyony et al. (1937), who named the elaborated movements "heteroeffector reflexes." A special term, *conditional-conditional reflexes,* has been introduced by Ivanov-Smolensky (1928) for newly learned motor reactions triggered by a conditional stimulus. The term emphasizes not only an afferent but also an efferent part of the reaction elaborated in the process of learning by establishing new conditional connections between and inside various motor structures.

A particular case of reorganization of a natural synergy (placing reaction in cats) is its triggering from unusual afferent input. Neuronal mechanisms of such reorganization have been studied by Kotlyar et al. (1983), Mayorov (1994), and others.

A well-known experimental model of motor reorganization has been changing handedness in rats by a forced practice (Peterson, 1934). Such a procedure requires inhibition of the previously preferred hand during food retrievals. After learning, the animals have used the other hand in free conditions for up to 270 days (Wentworth, 1942).

The retrieval of food from a narrow horizontal tube consists of a fast (ballistic) extension of an arm and its slower retraction. Bures et al. (1988) trained rats to increase the time of extension (i.e., to rearrange the ballistic phase). In another experiment of Bures et al. (1988), the rats were trained to modify rhythmic activity of a licking generator.

Several experimental models of learned inhibition of natural coordinations have been developed in dogs. The reorganization of natural coordination has been obtained by instrumental conditioning. Figure 7.1 shows two samples of elaborated motor reactions, including reorganized natural synergies. Figure 7.1b shows stages of elaboration of an alimentary instrumental reaction when the dog has to stay in contact with a food cup by lifting a forelimb during eating (Popova, 1970). An innate synergy (lowering the limb during lowering the head into the feeder) interferes with the reaction being learned (keeping the limb lifted during eating with lowered head). This results in synchronous oscillatory movements of the head and the limb (figure 7.1b; A). In the process of learning, the innate synergy is inhibited, and the dog becomes able to keep the limb lifted during eating (figure 7.1b; B, C). Another example, a precise avoidance reaction (Frolov, 1983), is represented in figure 7.1c. In the experimental procedure, if the dog lifts the limb above a "safety zone" in response to a conditional stimulus, it receives a continuous electrical stimulation of the limb, evoking a flexor reflex. To escape the pain, the dog has to inhibit the flexor reflex and lower the limb into the safety zone (figure 7.1c; A).

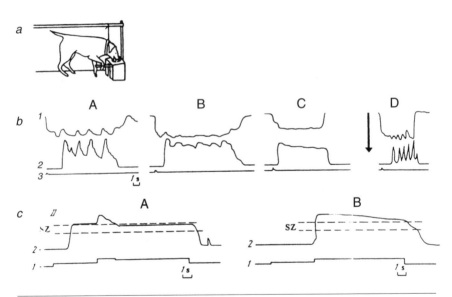

Figure 7.1   Two experimental models of rearrangement of innate coordination by learning in dogs. *(a)* Sketch of the experiment (see Pavlova and Alexandrov, 1992). *(b)* Alimentary instrumental reaction of lifting a cup of food and maintaining it during eating by a lifted forelimb. A, B, and C are sequential stages of training; D is after a lesion of the contralateral motor cortex; (1) head movement, (2) movement of the performing limb, (3) mark of the moment when the cup of food becomes available. *(c)* Precise avoidance reaction; A is after training; B is after a lesion of the contralateral motor cortex; (1) marks of conditioned (sound) and unconditioned (electrical stimulation of the limb) stimuli; (2) limb movement. SZ = safety zone.

Reprinted from Balezina et al. 1990.

Motor Cortex Is Involved in Motor Learning

During the last decade, classical conceptions concerning neural basis of motor learning have been revised because of a number of new data obtained by using modern approaches.

Functional Plasticity of the Motor Cortex

In the classical literature, the motor area of the cerebral cortex is known to control precision and fineness of motor actions, particularly isolated finger movements and their coordination, such as during a precision grip. After MCx or pyramidal tract lesions in primates, separate finger movements disappear and are replaced by a synergic flexion of all fingers. Strong somatotopic organization of the MCx is well known, and a conception suggesting that the output neurons of the MCx are "upper motoneurons" projecting point-to-point to the spinal motoneurons has been put forward (Phillips and Porter, 1977), although an overlap of many muscle representations has been shown.

Some studies showed that the point-to-point projections from the MCx are not stable and rigid. After partial lesions of the MCx, excitability of adjacent areas increased and their stimulation evoked movements previously evoked from the ablated part (Glees and Cole, 1950; Zhukov, 1895). During the last decade, intensive studies of the plasticity of the motor cortical representation have been carried out. The plasticity has been shown by various means. First, repetitive microstimulation of the MCx hand area in monkeys (Nudo et al., 1990) or microstimulation of the forelimb area combined with intracortical bicuculine injection in rats (Jacobs and Donoghue, 1991) resulted in an increase of the hand or forelimb representation areas. Also, microstimulation after MCx lesions in monkeys followed by intensive training of the appropriate limb (Nudo et al., 1996), after training without lesion (Milliken et al., 1992), after microstimulation combined with changes of a limb position (Sanes et al., 1992) or with passive movements (Humphrey et al., 1990), and after section of a peripheral nerve (Sanes and Donoghue, 1992) resulted in changes in the motor cortical map.

Some human studies have also shown plasticity of the motor cortical representation, as assessed by transcranial magnetic stimulation, after increased muscle activity (Braille reading) (Pascual-Leone et al., 1993) and during implicit or explicit learning of a definite sequence of key strikes by different fingers (Pascual-Leone et al., 1994). It has been shown that the plasticity of the map is NMDA-receptor dependent since an NMDA-antagonist MK-801 blocks plasticity induced by passive movements in rats (Qiu et al., 1990).

Another group of findings concerns long-term potentiation (LTP) in MCx (Asanuma, 1989; Donoghue et al., 1996). Changes in synaptic efficacy may provide the basis of reorganization in MCx during learning. Recently, new data have been revealed, showing that intrinsic horizontal neuronal connections in the primary motor cortex may be responsible for MCx plasticity. They show activity-dependent plasticity, and they modify in association with skill learning (Sanes and Donoghue, 2000).

Motor Cortex Is Activated During Motor Learning But Also During Skill Practice

Some earlier data have shown that the MCx deals with motor learning (Ioffe, 1973). Later, synaptogenesis and fos gene expression in the motor cortex have been shown after skill acquisition (Kleim et al., 1996). An activation of a set of brain structures, including contralateral MCx during different forms of motor learning, has been directly revealed by brain imaging techniques (PET and fMRI) (Grafton et al., 1995; Honda et al., 1998; Sadato et al., 1996; Seitz and Roland, 1992; van Mier et al., 1998). Interestingly, the digits and wrist area in the MCx increased in rats after motor skill learning but not after unskilled learning (Kleim et al., 1998). This would suggest a specific participation of the MCx in skill learning. In some brain imaging studies, however, the MCx activation was shown not to be specific for the process of learning since it was also observed during practice of previously learned movements (Jenkins et al., 1994; Karni et al., 1998). A detailed analysis showed that in some cases of learning, practice produces a shift in activity from one set of areas to another. This practice-related activation appears in the same hemisphere independently of the side of performance and may concern memory coding rather than task performance (van Mier et al., 1998).

Thus, the data show that the MCx has a functional plasticity and it participates in the process of motor learning. However, they do not show a specific function of the MCx in motor learning.

Motor Cortex Inhibits Inappropriate Motor Patterns During Motor Learning

What is the specific function of the motor cortex in motor learning? This question was studied in the following experiments.

Effects of Motor Cortex Lesions on Learned Coordinations

The motor cortex has been shown to play a crucial role in the inhibition of innate coordinations during their reorganization in the process of learning. After MCx lesions in the hemisphere contralateral to the performing limb, learned coordinations conflicting with natural ones disappear and the natural coordinations dominate (figure 7.1a; D, figure 7.1b; B). The limb movement disturbed after the surgery gradually recovers, but no learned coordination can be performed. In the case of the alimentary motor reaction, the motor pattern after MCx lesions is similar to that during the initial stage of training. The dog cannot keep the forelimb lifted while eating, and only oscillatory movements of the head and forelimb are possible; lowering the head into the feeder is accompanied by lowering the limb, and lifting the limb is accompanied by lifting the head. In the case of the avoidance reaction, the learned inhibition of the flexor reflex is severely disturbed after lesions of the contralateral motor cortex. In either paradigm, no recovery of the learned coordination can be observed even after 3 to 4 years of retraining. The same results can be obtained after a pyramidal section. In contrast, after surgery, the dog can maintain the limb lifted during eating if the food is presented from above (using a block or a high feeder) and the dog does not have to lower its head while lifting the limb (figure 7.2). Thus, during learning, a pattern of descending corticofugal signals was formed, providing inhibition of the natural coordination interfering with the learned one. After MCx lesions, the learned inhibitory influences disappeared and the natural coordination (lowering the limb while lowering the head) dominated. During eating with the head lifted, the natural coordination did not interfere with the movement, and it was possible to keep the limb lifted.

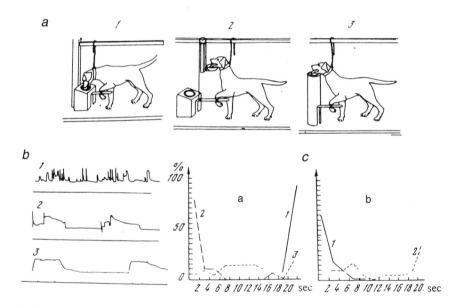

Figure 7.2   Alimentary instrumental reactions in pyramidotomized dogs with different head positions during eating. *(a)* Variants of the experiment: eating with (1) lowered and (2, 3) lifted head. *(b)* Motor reactions with (1) lowered and (2, 3) lifted head. *(c)* Period when the limb is lifted during eating after pyramidotomy performed (a) after or (b) before the training. In (a) (1) before the surgery, (2) after the surgery with lowered head, (3) after the surgery with lifted head. In (b) (1) with lowered head, (2) with lifted head. Abscissa: time when the limb is lifted during eating; ordinate: percentage of the corresponding reactions. Note that, after section of pyramids, the limb remains lifted longer during eating with a lifted head compared to eating with a lowered head but still for much less time than in a healthy animal.

Reprinted from Ioffe 1973.

A particular question concerns the neural basis of the natural head-forelimb coordination. A role of the neck and vestibular reflexes has been specifically tested. However, neither vestibulectomy nor neck deafferentation resulted in any changes in the natural coordination after MCx lesions (Pavlova, 1996a). Perhaps the natural coordination has a more complicated organization. Another interesting finding is that the previously described disturbances of the learned head-forelimb coordination can be observed after selective lesions of the forelimb representation in MCx sparing the head representation (Pavlova, 1996b). Perhaps the new pattern of neuronal connections providing inhibition of the natural head-forelimb synergy in the process of learning is formed only in the forelimb MCx area.

Thus, a particular function of the motor cortex in motor learning is inhibition of reflexes and synergies that interfere with a movement being learned.

Specificity of Motor Cortex in Inhibition of Inappropriate Motor Patterns

Is the program for learned inhibition of inappropriate coordinations formed in the motor cortex, or is the MCx only relayed from associative cortical areas or the cerebellum? After lesions of parietal associative cortex, no disturbances of the learned head-forelimb coordination have been observed. Similar results have been obtained after lesions of premotor cortical areas. Combined lesions of parietal and premotor associative cortex, however, resulted in disturbances of the learned coordination that were very similar to that observed after MCx lesions (figure 7.3). In contrast to the MCx lesions, the disturbances disappeared after 3 to 4 weeks of retraining (Pavlova et al., 1986). Similar results have been obtained after lesions of the cerebellar nuclei (Balezina and Mats, 1995). In all cases, the learned coordination could be recovered.

Thus, the inhibition of inappropriate synergies and coordination during motor learning is a specific function of the motor cortex.

Motor Cortex Participates in Control of Postural Adjustments: Animal Experiments

The previously mentioned data concern functions of the motor cortex in the control of limb movements. However, limb movements are usually accompanied by postural adjustments, realized by proximal and axial musculature. Until relatively recently, the motor cortex was assumed not to be involved in the control of postural adjustments preceding and accompanying limb movements. That opinion was based, in particular, on data obtained by Shumilina (1949) and Koryakin (1958), who observed dramatic disturbances of a learned limb movement, but not postural adjustments, after bilateral MCx lesions. In other experiments, stimulation of the motor cortex caused limb lifting without postural adjustment, which resulted in the

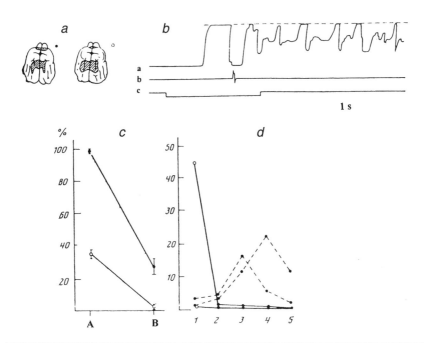

Figure 7.3   Effect of combined bilateral lesions of parietal and premotor cortex on performance of the alimentary motor reaction (maintaining the limb lifted during eating with lowered head). *(a)* Schemes of the lesions in two dogs. *(b)* An example of the performance of the previously learned motor reaction after surgery: (a) limb movement, (b) mark of food presentation, (c) duration of the sound conditional stimulus. *(c)* Mean amplitude of the lifted limb in percentage of the control level (broken line in *(b)*) (A) before and (B) after the surgery in the two dogs with the lesions presented in *(a)*. *(d)* Frequency of the limb fluctuations with different amplitudes in the period of a lifted limb during eating in the same dogs before (solid lines) and after (broken lines) the surgery. Abscissa: amplitude levels of the fluctuations (1, minimal; 5, maximal); ordinate: number of the corresponding fluctuations per minute.
Reprinted from Pavlova, Balezina, and Ioffe 1986.

animal falling (Konorski, 1967; Tarnecki, 1962; Thomas, 1971; Wagner et al., 1967). When an intracortical MCx stimulation of much lower intensity was used (Gahery and Nieoullon, 1978), no falling occurred, but the initial reaction was similar (decrease of the support force of the appropriate limb). Short-latency electromyographic (EMG) changes in the ipsilateral limbs were also a reason for the opinion that the postural adjustment might be activated by collaterals of the pyramidal tract in the brainstem or at the spinal level (Massion, 1992). An analysis of the center of pressure trajectory showed, however, that it was corrected only about 100-150 ms after onset of MCx stimulation (Ioffe et al., 1982), which suggested a reflex correction. Perhaps, both mechanisms coexist.

Various structures have been believed to be responsible for the control of postural adjustments, particularly the basal ganglia (Martin, 1967; Shapovalova et al., 1984), the cerebellum (Massion, 1979), and the brainstem reticular formation (Gorska et al., 1996; Ioffe, 1991; Koryakin, 1958; Shumilina, 1949). The red nucleus has been shown to influence postural adjustment latency (Burlachkova and Ioffe, 1979). However, as early as the 1930s, the postural placing reaction was found to be under cortical control (Bard, 1933). Subsequent studies revealed that the motor cortex also takes part in the control of postural reactions that provide a shift of the center of mass (COM) (Birjukova et al., 1989; Ioffe, 1999; Ioffe et al., 1988; Massion, 1979). Motor cortex influences on postural control have been found to be bilateral. Unilateral MCx lesions resulted in a latency increase of the support force response in all four limbs, although the contralateral limb movement latency increased much more (Burlachkova and Ioffe, 1979). After motor cortex lesions, a previously ballistic COM displacement is realized in a step-by-step manner, under permanent afferent control (Birjukova et al., 1989; Ioffe et al., 1988). Figure 7.4a shows changes of the COM displacement, velocity, and acceleration in an intact animal, and figure 7.4b shows changes after motor cortex lesions. One can see that the COM shift was ballistic before the surgery, but after the lesion, many peaks of acceleration and velocity were present. Deceleration of the center of mass (i.e., stabilization of the body and equilibrium maintenance in the final position) is difficult as well. However, the disturbances of postural adjustment may be compensated after intensive retraining. The bilateral effect of unilateral MCx lesions suggests modulatory cortical influence on (perhaps brainstem) structures controlling postural adjustments. Possibly, the motor cortex coordinates and integrates motor programs for limb movement and postural adjustment.

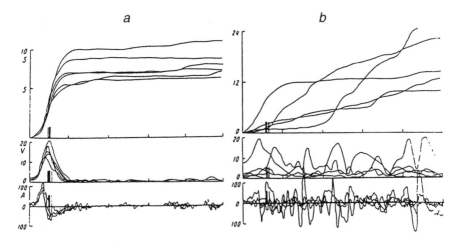

Figure 7.4   Changes of the horizontal displacement (S, cm), velocity (V, cm/s), and acceleration (A, cm/s/s) of the center of gravity during several trials. *(a)* A well-trained avoidance reaction (*n* = 6). *(b)* After lesion of the contralateral motor cortex, before retraining (*n* = 5). The abscissa shows time (s). The vertical bars correspond to the limb liftoff.

Reprinted from Alexandrov, Vasilyeva, Ioffe, et al. 1991.

Learned Reorganization of a Pattern of Postural Adjustment: Role of the Motor Cortex in Inhibition of a Natural Pattern

Some animal models have been developed of reorganization of a postural adjustment pattern based on suppression of a natural pattern. One such model is a reorganization of the so-called diagonal pattern of postural adjustment accompanying a limb movement. Usually, the limb diagonally opposite to the lifted one is unloaded, whereas the other pair of diagonal limbs is loaded (figure 7.5a). Through special training, it is possible to train an alternative, the so-called unilateral (ipsilateral) pattern of postural adjustment: the limb ipsilateral to the lifted one is unloaded, whereas the other pair of ipsilateral limbs is loaded (figure 7.5b). This pattern may

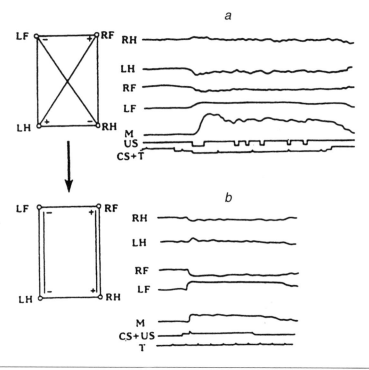

Figure 7.5   Rearrangement of a "diagonal" pattern of postural adjustment accompanying an avoidance reaction into a "unilateral" one. *(a)* The diagonal pattern of postural adjustment (simultaneous unloading versus loading in pairs of diagonal limbs). *(b)* The unilateral pattern of postural adjustment (simultaneous unloading versus loading in pairs of ipsilateral limbs). On the left are schemes of the support force changes; on the right, fragments of the recordings: LF, RF, LH, RH, force traces of the left and right forelimbs and left and right hindlimbs, respectively; M: trace of the limb movement; CS, US: marks of conditioned and unconditioned stimuli; T: time marks, s.

be obtained using avoidance reflex in which electrical stimulation is applied to two ipsilateral limbs in dogs (figure 7.5, left forelimb and left hindlimb). The dog can avoid or escape the stimulation of one limb (figure 7.5, left forelimb) by lifting and holding it above a certain level (usually 5-7 cm [2.0-2.75 in.]) for a definite time (usually 4-5 s), whereas the stimulation of the other (ipsilateral) limb may be avoided or escaped by decreasing the support force of the limb by 10% of the initial level. As a result of such training, the diagonal pattern of the postural adjustment is suppressed and replaced by a unilateral one.

The degree of diagonality was expressed by a formula (Gahery et al., 1980) where the maximum of the diagonality coefficient (D) (standing on two diagonal limbs) corresponded to 1. The dynamics of D during reorganization of the diagonal pattern of postural adjustment into the unilateral one are represented in figure 7.6a.

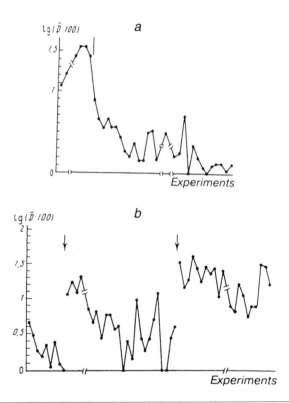

Figure 7.6 *(a)* Changes in the diagonality coefficient in the course of rearrangement of the "diagonal" pattern of postural adjustment into the unilateral one. *(b)* After sequential lesions of the motor cortex in the contralateral (left arrow) and ipsilateral (right arrow) hemispheres. Abscissa: sequential sessions; ordinate: decimal logarithm of the diagonality coefficient expressed in percentage. Vertical bar in *(a)* corresponds to the start of the postural pattern rearrangement.

Reprinted from Ioffe, Ivanova, Frolov, et al. 1988.

A lesion of the motor cortex contralateral to the stimulated ("active") limb results in temporary disturbances of the reorganized pattern of postural adjustment, which may be compensated by 3 to 4 weeks of retraining (figure 7.6b). However, a subsequent lesion of the motor cortex in the other hemisphere causes a stable disappearance of the learned postural pattern, and only the diagonal pattern of postural adjustment is observed. No recovery of the unilateral pattern can be obtained by retraining (figure 7.6b). Thus, the motor cortex apparently inhibits inappropriate postural coordinations during their rearrangement. However, in contrast to the control of limb movements, effects of the motor cortex during reorganization of postural coordinations are bilateral. This is understandable, taking into account that the pattern of postural adjustment is bilateral as well. Perhaps, in the process of learning, MCx modulates the activity of brainstem generators, which are responsible for organization of the pattern of postural adjustment. As a result, an innate postural pattern can be inhibited and substituted for by the learned one.

Human Studies: Coordination Between Posture and Movement in a Bimanual Load-Lifting Task

Earlier, we posited that the MCx is important for acquisition of a new coordination between posture and movement in animals. A convenient model for studying this problem in humans is a bimanual load-lifting task. When a load supported by one hand is lifted off by the other hand (voluntary unloading), the position of the "postural" forearm remains almost unchanged, although the disturbance resulting from the unloading might be expected to trigger an upward forearm movement for mechanical reasons (as is usually observed during imposed unloading; see figure 7.7). This lack of forearm movement during voluntary unloading is due to an anticipatory postural adjustment associated with the lifting movement, which consists of an inhibition of the postural forearm flexors starting before the onset of unloading (Dufosse et al., 1985; Hugon et al., 1982; Massion, 1992; Massion et al., 1998). This postural adjustment is based on a feedforward control and minimizes the forearm position disturbance induced by the unloading. This coordination between posture and movement is developed in childhood. Some degree of anticipation in this task is present in 3- to 4-year-old children (Schmitz et al., 1999). In adults, this coordination is very usual and is often called the "barman effect." Here, we will use "natural coordination."

In some artificial situations of bimanual unloading, the anticipatory postural adjustment is absent. This can be observed, for example, when the unloading of one forearm is triggered by lifting off another load from a strain-gauge platform by the other forearm (Paulignan et al., 1989). This is also true when the voluntary movement triggering the unloading is a leg movement (Forget and Lamarre, 1990). In these cases, the unloading evokes an elbow rotation. After several repetitions of the unloading procedure, however, the anticipatory postural adjustment (decrease of the flexor activity of the "postural" forearm before the unloading onset)

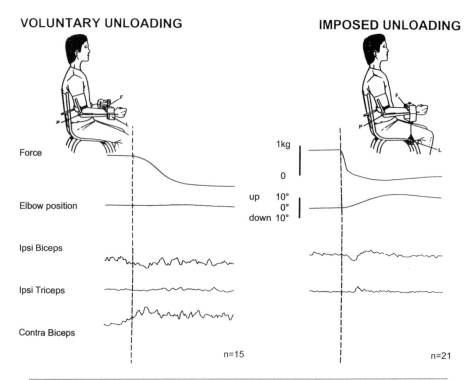

Figure 7.7   Changes in the forearm position during voluntary and imposed unloading. The averaged traces, from top to bottom: force, angular displacement of the forearm, and integrated EMG activities of the biceps and triceps muscles of the postural arm and the biceps of the moving arm (in the situation of voluntary unloading). Vertical line corresponds to the onset of unloading.

Reprinted from Viallet, Massion, Massarino, et al. 1987.

arises, and the maximal amplitude and maximal velocity of the "postural" forearm flexion decrease significantly. Thus, the acquisition of a new coordination between posture and movement takes place, resulting in minimizing elbow rotation during forearm unloading (Paulignan et al., 1989). The elbow stiffness of the postural forearm has been found to be increased during the acquisition (Biryukova et al., 1999). Figure 7.8 shows changes of the maximal amplitude of forearm flexion in three sequential series of such a learning in two healthy subjects (20 trials in each series; the interval between the series is about 5 minutes). The reduction of the amplitude can be seen both inside the series and between them. The learning seems to go faster in the first series. In some subjects, the learning could be observed mainly in the first series, as is seen in figure 7.8c. The learning has been shown to be unilateral, with no transfer of the learned reduction of the maximal amplitude and maximal velocity after changing the postural and moving forearms (Ioffe et al., 1996).

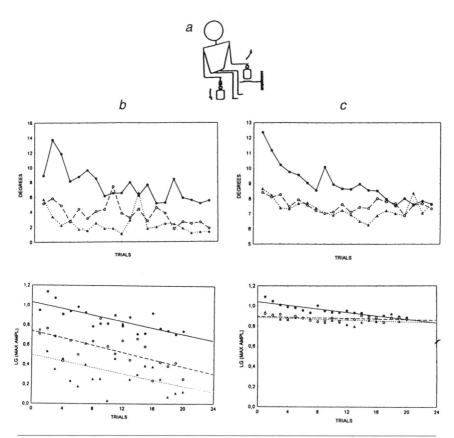

Figure 7.8   Changes in the maximal amplitude of forearm flexion during acquisition of a new coordination between posture and movement in the bimanual load-lifting task. *(a)* Sketch of the experimental paradigm. *(b, c)* Amplitude of the forearm flexion (top panels) and the corresponding regression lines (bottom panels) in two subjects in the first (filled circles, solid lines), second (open circles, broken lines), and third (triangles, dotted lines) experimental series. Abscissa: sequential trials; ordinate: maximal amplitude of the elbow flexion, degrees (top panels) or decimal logarithm of the maximal amplitude (bottom panels).

Deficit of Acquisition of a New Coordination Between Posture and Movement in Parkinsonian and Hemiparetic Patients: A Specific Role of Capsular Pathways in Learning a New Coordination

The acquisition of the new coordination between posture and movement in the bimanual load-lifting task has been studied in a group of healthy subjects ($n = 8$) aged 57-65, a group of parkinsonian patients ($n = 10$), and a group of patients after cerebrovascular accidents accompanied by asymmetrical motor disturbances

(n = 13). The last group consisted of a group of hemiparetic patients with capsular lesions (n = 7) and a group of patients with lesions involving frontal or parietal cortex, midbrain tegmentum, and so on but sparing the internal capsule (n = 6). During the experiments, the subjects were sitting on a chair holding a forearm horizontally with the elbow angle at about 90°. A load of 1 kg was fixed to the wrist of the postural forearm by an electromagnet, which could be released by lifting another weight of 1 kg by the other forearm from a strain-gauge platform or by a computer command given by the experimenter. The protocol of the investigation was the same in all the groups and included a control series of 20 trials of unexpected unloading imposed by the experimenter, two sequential learning series of 20 trials, a second control series, a third learning series, and a series of 20 trials of voluntary unloading when the subject lifted off the weight from the wrist of the postural forearm using the other arm. Taking into account the absence of transfer of the learned reduction of the amplitude and velocity of forearm flexion, both affected and nonaffected forearms (contralateral and ipsilateral to the lesioned hemisphere) were used as a postural forearm on two separate experimental days (Massion et al., 1999).

Figure 7.9 shows mean values of the maximal amplitude of forearm flexion in the process of learning in different groups of subjects. Several differences among the

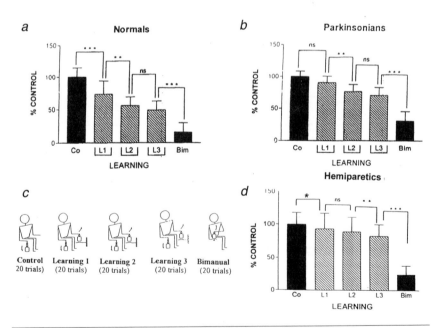

Figure 7.9   Maximal amplitude of the postural forearm flexion (mean ± SD) as percent of the mean value in control (Co), three sequential learning series (L1, L2, L3), and in natural coordination (Bim) *(a)* in healthy subjects, *(b)* parkinsonian patients, and *(d)* hemiparetic patients with capsular lesions contralateral to the postural arm. *(c)* The experimental paradigms. Statistical differences among the series are marked by * (p < .05), ** (p < .01), and *** (p < .001); ns: nonsignificant differences.

groups can be noticed. First, in the group of healthy subjects, the acquisition is more pronounced in the first series, whereas in the other two groups, it develops more slowly. Second, the maximal level of acquisition (in the third series) is about 50% of the control level (48.3 ± 18.4%) in the group of healthy subjects, whereas it is much less in the group of parkinsonian patients (68.6 ± 17.1%) and hemiparetic patients (83.8 ± 18.9%). Thus, the acquisition is disturbed in parkinsonian and particularly in hemiparetic patients, although even in this group some degree of learning exists. The natural coordination is also disturbed both in parkinsonian and hemiparetic patients. The disturbances of the natural coordination were described elsewhere (Viallet et al., 1987, 1992). However, in contrast to the deficit in acquisition, the disturbances of the natural coordination seem to be maximal in the parkinsonian but not in the hemiparetic group.

In hemiparetic patients, the learning is disturbed both in the contralateral (to the lesioned hemisphere) and ipsilateral forearms. However, the deficit is significantly larger on the contralateral side (figure 7.10). When the ipsilateral forearm is used as a postural one, there is a considerable reduction of the maximal amplitude already in the first series, in contrast to the contralateral side and to parkinsonian patients, where the learning is markedly delayed.

Figure 7.11 represents the maximal level of acquisition versus natural coordination in all the tested groups. The greatest deficit in acquisition is found in the hemiparetic patients on the affected side (contralateral to the lesioned hemisphere). Interestingly, the degree of acquisition disturbances did not differ among the groups of parkinsonian patients, patients with noncapsular lesions, and the nonaffected side (ipsilateral to the lesioned hemisphere) of hemiparetic patients.

The results suggest a specific role of the descending pathways going in the internal capsule in acquisition of a new coordination between posture and movement

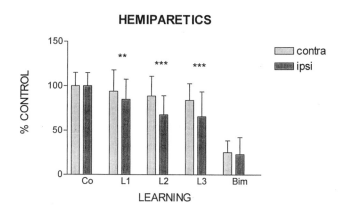

Figure 7.10   Maximal amplitude of the postural forearm flexion in percentage of mean control value in hemiparetic patients on the affected (contralateral to the lesioned hemisphere) and nonaffected (ipsilateral to the lesioned hemisphere) sides. The designations are as in figure 7.9.

ACQUISITION vs. NATURAL COORDINATION

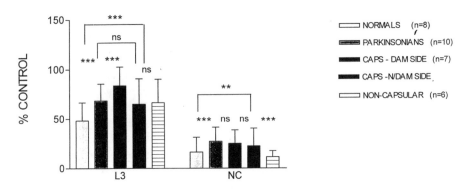

Figure 7.11   Maximal level of acquisition of a new coordination between posture and movement in the bimanual load-lifting task (maximal amplitude of the forearm unloading movement in percentage of control) in the third learning series (L3) and maximal amplitude during voluntary unloading (natural coordination, NC) in all the investigated groups. The greatest deficit is in learning of patients with capsular lesions on the affected side (deficit in specific descending commands for learning a new coordination). In contrast, the absence of significant differences among the groups in the natural coordination points to a nonspecific function of the studied structures in the control of a well-learned coordination.

in the bimanual load-lifting task. In addition, generalized (so-called nonspecific) facilitatory input from the nigrostriatal system and other brain structures may influence acquisition at the spinal level.

Surprisingly, the disturbances of the natural coordination did not significantly differ between hemiparetic and parkinsonian patients, although they were somewhat more pronounced in parkinsonian patients; see figure 7.9. This suggests a nonspecific descending effect of the studied structures on the performance of a well-learned ("natural") posturo-kinetic coordination.

What is the specific role of the internal capsule in acquisition of the new coordination between posture and movement? The observed anticipatory postural adjustment consisted of an inhibition of the elbow flexor activity before the unloading. This inhibition might be a result of the activity of the capsular descending (perhaps pyramidal) pathways. Assuming that this activity is generated by the motor cortex, the earlier mentioned role of the MCx in inhibiting inappropriate motor patterns during motor learning might be applicable to the bimanual load-lifting task as well.

The specific capsular inhibition of the elbow flexor activity before the unloading develops in ontogenesis. Very young children use various EMG patterns of anticipatory postural adjustment preceding unloading. Initially, they use mainly co-contractions and progressively select the inhibitory pattern in the flexor. Learning seems to involve the inhibition of these immature patterns, rather than an innate one, because the large repertoire of EMG patterns seems to coexist up to the age of 8 years (Schmitz et al., 1999).

Conclusion

The motor cortex not only controls precision and fineness of movements but is involved in the process of motor learning as well. A specific function of the motor cortex in motor learning is inhibition of synergies and coordinations conflicting with the movements being learned. In the process of learning, new neuronal interconnections in the motor cortex and descending commands are organized that provide a pattern of a new movement and inhibition of inappropriate motor patterns. After a motor cortex lesion, the inhibition disappears and natural coordinations dominate, preventing performance of the learned movements.

The motor cortex controls not only limb movements but also postural adjustments. In animals, it provides a learned inhibition of an innate diagonal pattern of postural adjustment. In humans, descending pathways passing through the internal capsule specifically contribute to acquisition of a new coordination between posture and movement in a bimanual load-lifting task, possibly dealing with anticipatory postural adjustment (inhibition of muscle activity of elbow flexors prior to unloading). In hemiparetic patients with capsular lesions, the acquisition is dramatically disturbed in the contralateral forearm, whereas the deficit on the ipsilateral side is significantly smaller and does not differ from the acquisition deficit in parkinsonian patients and patients with brain lesions sparing the internal capsule (nonspecific influence).

Interestingly, the disturbances of a natural coordination between posture and movement (voluntary unloading) did not significantly differ among parkinsonian, hemiparetic, and other groups of patients, revealing an absence of a specific role of any of the lesioned structures in the performance. The data suggest a specific descending influence via the internal capsule on the acquisition of a new coordination between posture and movement in the bimanual load-lifting task but not on the performance of a well-learned coordination. This seems contrary to animal experiments showing that the motor cortex and pyramidal pathways are required not only for the acquisition of a new coordination but also for the performance of a learned coordination based on the inhibition of an innate coordination. In humans, the anticipatory postural adjustment (inhibition of the forearm flexor activity prior to the voluntary unloading) learned in early ontogenesis might be performed without direct corticospinal influences.

Acknowledgments

The authors thank Nicolas Martin for creating the experimental software, Roselyne Aurenty for technical assistance, Milan Zedka for reading and correcting the manuscript, and Zhanna Shuranova for help in preparation of the manuscript. The work was supported by the Russian Foundation of Basic Research, projects No. 99-04-48885 and 01-04-49296, and by Russian Humanitarian Scientific Foundation, project No. 00-06-00242a.

References

Alexandrov, A.V., Vasilyeva, O.N., Ioffe, M.E., and Frolov, A.A. (1991) Some modes of description of different patterns of postural adjustment during motor learning in dogs. *Zh Vyssh Nerv Deiat* 41: 937-947.

Asanuma, H. (1989) *The motor cortex.* New York: Plenum Press.

Balezina, N.P., and Mats, V.N. (1995) Participation of the cerebellar nuclei in elaboration and realization of a learned motor coordination in dogs. In Fanardjian, V.V. (Ed.), *Cerebellum and brainstem structures* (in Russian), pp. 184-191. Erevan: Gitutyun.

Balezina, N.P., Varga, M.E., Vasilyeva, O.N., Ivanova, N.G., Ioffe, M.E., Pavlova, O.G., and Frolov, A.G. (1990) A study of mechanisms of reorganisation of motor coordinations during motor learning. In Airapetyants, M.G. (Ed.), *Brain and behaviour* (in Russian), pp.105-119. Moscow: Nauka.

Bard, P. (1933) Studies on the cerebral cortex: 1. Localized control of placing and hopping reactions in the cat and their normal management by small cortical lesions. *Arch Neurol and Psychiatry* 30: 40-74.

Bernstein, N.A. (1967) *The coordination and regulation of movements.* Oxford-London-New York-Toronto-Sydney-Paris: Pergamon Press.

Birjukova, E.V., Dufosse, M., Frolov, A.A., Ioffe, M., and Massion, J. (1989) Role of the sensorimotor cortex in postural adjustment accompanying a conditioned paw lift in the standing cat. *Exp Brain Res* 78: 588-596.

Biryukova, E.V., Roschin, V.Y., Frolov, A.A., Ioffe, M.E., Massion, J., and Dufosse, M. (1999) Forearm postural control during unloading: Anticipatory changes in elbow stiffness. *Exp Brain Res* 129: 110-120.

Bures, J., Buresova, O., and Krivanek, J. (1988) *Brain and behavior. Paradigms for research in neural mechanisms.* Prague: Academia.

Burlachkova, N.I., and Ioffe, M.E. (1979) The analysis of the postural adjustment accompanying a local movement. *Agressologie* 20: 141-142.

Donoghue, J.P., Hess, G., and Sanes, J.N. (1996) Substrates and mechanisms for learning in motor cortex. In Bloedel, J.M., Ebner, T.J., and Wise, S.P. (Eds.), *The acquisition of motor behavior in vertebrates,* pp. 363-386. Cambridge, MA: MIT Press.

Dufosse, M., Hugon, M., and Massion, J. (1985) Postural forearm changes induced by predictable in time or voluntary triggered unloading in man. *Exp Brain Res* 60: 330-334.

Fitts, P.M. (1954) The information capacity of the human motor system in controlling the amplitude of movement. *J Exp Psychol* 67: 381-391.

Forget, R., and Lamarre, Y. (1990) Anticipatory postural adjustment in the absence of normal peripheral feedback. *Brain Res* 508: 176-179.

Frolov, A.G. (1983) The effect of instrumentalization of inborn reaction on its transformation into contrary directed escape response in dogs and the problem of reinforcement. *Acta Neurobiol Exp* 43: 1-14.

Gahery, Y., Ioffe, M.E., Massion, J., and Polit, A. (1980) The postural support of movement in cat and dog. *Acta Neurobiol Exp* 40: 741-756.

Gahery, Y., and Nieoullon, A. (1978) Postural and kinetic co-ordination following cortical stimuli induce flexion movements in cat's limbs. *Brain Res* 149: 25-37.

Glees, P., and Cole, J. (1950) Recovery of skilled motor functions after small repeated lesions of motor cortex in macaque. *J Neurophysiol* 13(2): 137-148.

Gorska, T., Ioffe, M., Zmyslowski, W., Bem, T., Majczynski, H., and Mats, V.N. (1996) Unrestrained walking in cats with medial pontine lesions. *Brain Res Bull* 38: 297-304.

Grafton, S.T., Hazeline, E., and Ivry, R. (1995) Functional mapping of sequence learning in normal humans. *J Cognitive Neurosci* 7: 497-510.

Hallett, M., Pascual-Leone, A., and Topka, H. (1996) Adaptation and skill learning: Evidence for different neural substrates. In Bloedel, J.M., Ebner, T.J., and Wise, S.P. (Eds.), *The acquisition of motor behavior in vertebrates,* pp. 289-302. Cambridge, MA: MIT Press.

Honda, M., Deiber, M.-P., Ibanez, V., Pascual-Leone, A., Zhuang, P., and Hallett, M. (1998) Dynamic cortical involvement in implicit and explicit motor sequence learning. A PET study. *Brain* 121: 2159-2173.

Hugon, M., Massion, J., and Wiesendanger, M. (1982) Anticipatory postural changes induced by active unloading and comparison with passive unloading in man. *Pflugers Arch* 393: 292-296.

Humphrey, D.R., Qiu, X.Q., Clavel, P., and O'Donoghue, D.L. (1990) Changes in forelimb motor representation in rodent cortex induced by passive movements. *Society for Neuroscience Abstracts* 16: 422.

Ioffe, M.E. (1973) Pyramidal influences in establishment of new motor coordinations in dogs. *Physiol Behav* 11: 145-153.

Ioffe, M.E. (1991) *Mechanisms of motor learning* (in Russian). Moscow: Nauka.

Ioffe, M.E. (1999) The motor cortex inhibits synergies interfering with a learned movement: Reorganization of postural coordination in dog. In Ivanitsky, A.M., and Balabau, P.M. (Eds.), *Complex brain function: Conceptual advances in Russian neuroscience,* pp. 289-300. London: Harwood.

Ioffe, M.E., Frolov, A.A., Gahery, Y., Frolov, A.G., Coulmance, M., and Davydov, V.I. (1982) Biomechanical study of the mechanisms of postural adjustment accompanying learned and induced limb movements in cats and dogs. *Acta Neurophysiol Exp* 42: 469-482.

Ioffe, M.E., Ivanova, N.G., Frolov, A.A., Birjukova, E.V., and Kiseljova, N.V. (1988) On the role of motor cortex in the learned rearrangement of postural coordinations. In Gurfinkel, V.S., Ioffe, M.E., Massion, J., and Roll, J.-P. (Eds.), *Stance and motion: Facts and concepts,* pp. 213-226. New York: Plenum Press.

Ioffe, M., Massion, J., Gantchev, N., Dufosse, M., and Kulikov, M.A. (1996) Coordination between posture and movement in a bimanual load-lifting task: Is there a transfer? *Exp Brain Res* 109: 450-456.

Ivanov-Smolensky, A.G. (1928) Basic forms of conditional and unconditional human activity and their anatomical basis. *Zh Nevropatol Psykhiatr* 21: 229-248.

Jacobs, K.M., and Donoghue, J.P. (1991) Reshaping the cortical motor map by unmasking latent intracortical connections. *Science* 251: 944-947.

Jaric, S., and Latash, M.L. (1998) Learning a motor task involving obstacles by a multi-joint, redundant limb: Two synergies within one movement. *J Electromyogr Kinesiol* 8: 169-176.

Jenkins, I.N., Brooks, D.J., Nixon, P.D., Frackowiak, R.S.J., and Passingham, R.E. (1994) Motor sequence learning: A study with positron emission tomography. *J Neurosci* 14: 3775-3790.

Karni, A., Meyer, G., Jezzard, P., Rey-Hipolito, C., Adams, M., Turner, R., and Ungerleider, L.G. (1998) The acquisition of skilled motor performance: Fast and slow experience-driven changes in primary motor cortex. *Proc Natl Acad Sci USA* 95: 861-868.

Kleim, J.A., Barbay, S., and Nudo, R.J. (1998) Functional reorganization of the rat motor cortex following motor skill learning. *J Neurophysiol* 80: 3321-3325.

Kleim, J.A., Lissing, E., Schwartz, E.R., Comery, T.A., and Greenough, W.T. (1996) Synaptogenesis and Fos expression in the motor cortex of the adult rat after complex motor skill acquisition. *J Neurosci* 16: 4529-4535.

Konorski, J. (1967) *Integrative activity of the brain.* Chicago: University of Chicago Press.

Koryakin, M.F. (1958) On structure of positional excitations in conditional defensive reflex in the dog. *Fiziol Zh SSSR* 44: 393-403.

Kotlyar, B.I., Mayorov, V.I., Timofeeva, N.O., and Shul'govsky, V.V. (1983) *Neuronal organization of the conditional reflex behaviour* (in Russian). Moscow: Nauka.

Martin, J.P. (1967) *The basal ganglia and posture.* London: Pitman.

Massion, J. (1979) Role of the motor cortex in the postural adjustment associated with movements. In Asanuma, H., and Wilson, V.J. (Eds.), *Integration in the nervous system,* pp. 239-260. Tokyo: Igaku-Shoin.

Massion, J. (1992) Movement, posture and equilibrium: Interaction and coordination. *Prog Neurobiol* 38: 35-56.

Massion, J., Alexandrov, A., and Vernazza, S. (1998) Coordinated control of posture and movement: Respective role of motor memory and external constraints. In Latash, M.L. (Ed.), *Progress in motor control*, pp. 127-150. Champaign, IL: Human Kinetics.

Massion, J., Ioffe, M., Schmitz, C., Viallet, F., and Gantcheva, R. (1999) Acquisition of anticipatory postural adjustments in a bimanual load lifting task: Normal and pathological aspects. *Exp Brain Res Suppl* 128: 229-235.

Mayorov, V.I. (1994) Mechanisms of development of neuronal reactions in cat's motor cortex associated with triggering the conditioned placing reaction: A hypothesis. *Zh Vyssh Nerv Deiat* 44: 963.

Milliken, G., Nudo, R., Grenda, R., Jenkins, W.M., and Merzenich, M.M. (1992) Expansion of distal forelimb representations in primary motor cortex of adult squirrel monkeys following motor training. *Soc Neurosci Abstr* 18: 214.

Nudo, R.J., Jenkins, W.M., and Merzenich, M.M. (1990) Repetitive microstimulation alters the cortical representation of movements in adult rats. *Somatosens Mot Res* 7: 463-483.

Nudo, R.J., Wise, B.M., Fuentes, F.S., and Milliken, G.W. (1996) Neural substrates for the effects of rehabilitative training of motor recovery after ischemic infarct. *Science* 272: 1791-1794.

Pascual-Leone, A., Cammarota, A., Wasserman, E.M., Brasil-Neto, J.P., Cohen, L.G., and Hallett, M. (1993) Modulation of motor cortical outputs to the reading hand of Braille readers. *Ann Neurol* 34: 33-37.

Pascual-Leone, A., Grafman, J., and Hallett, M. (1994) Modulation of cortical motor output maps during development of implicit and explicit learning. *Science* 263: 1287-1289.

Paulignan, Y., Dufosse, M., Hugon, M., and Massion, J. (1989) Acquisition of co-ordination between posture and movement in a bimanual task. *Exp Brain Res* 77: 337-348.

Pavlova, O.G. (1996a) Role of vestibular and neck reflexes in the innate head-forelimb movement coordination in dogs. In Stuart, D.G. (Ed.), *Motor control-VII*, pp. 187-189. Tucson, AZ: Motor Control Press.

Pavlova, O.G. (1996b) Role of the motor cortex projection areas in reorganization of the innate head-forelimb coordination in dogs. In Gantchev, G.N., Gurfinkel, V.S., Stuart, D., and Wiesendanger, M. (Eds.), *Motor control-VIII*, pp. 164-167. Sofia: Academic Press House.

Pavlova, O.G., and Alexandrov, A.V. (1992) Head-forelimb movement coordination and its rearrangement by learning in the dog. The role of motor cortex. In Berthoz, A., Graf, W., and Vidal, P.P. (Eds.), *The head-neck sensory-motor system*, pp. 591-596. New York: Oxford University Press.

Pavlova, O.G., Balezina, N.P., and Ioffe, M.E. (1986) Disorganization of a learned coordination after combined lesion of parietal and premotor associative areas in dogs. *Zh Vyssh Nerv Deiat* 36: 450-459.

Peterson, G.M. (1934) Mechanisms of handedness in rats. *J Comp Psychol Monogr* 9: 1-67.

Phillips, C.G., and Porter, R. (1977) *Corticospinal neurons: Their role in movement*. London: Academic Press.

Popova, E.I. (1970) *Instrumental motor reflexes in aspect of the theory of conditional reflexes*. D. Sci. Thesis (in Russian). Moscow: Institute of Higher Nervous Activity and Neurophysiology.

Qiu, X.Q., O'Donoghue, D.L., and Humphrey, D.R. (1990). NMDA-antagonist (MK-801) blocks plasticity of motor cortex maps induced by passive limb movement. *Society for Neuroscience Abstracts* 16: 422.

Sadato, N., Ibanez, V., Deiber, M-P., Campbell, J., Leonardo, M., and Hallett, M. (1996) Frequency-dependent changes of regional blood flow during finger movements. *J Cereb Blood Flow Metab* 16: 23-33.

Sanes, J., and Donoghue, J. (1992) Organization and adaptability of muscle representations in primary motor cortex. *Exp Brain Res Suppl* 22: 103-127.

Sanes, J., and Donoghue, J. (2000) Plasticity and primary motor cortex. *Ann Rev Neurosci* 23: 393-415.

Sanes, J.N., Wang, J., and Donoghue, J.P. (1992). Immediate and delayed changes of rat cortical output representation with new forelimb configurations. *Cereb Cortex* 2: 141-152.

Schmitz, C., Martin, N., and Assaiante, C. (1999) Development of anticipatory postural adjustments in a bimanual load-lifting task in children. *Exp Brain Res* 126: 200-204.

Seitz, R.J., and Roland, P.E. (1992) Learning of sequential finger movements in man: A combined kinematic and positron emission tomography (PET) study. *Eur J Neurosci* 4: 154-165.

Shapovalova, K.B., Yakunin, I.V., and Boiko, M.I. (1984) Participation of head of caudate nucleus in mechanisms of conditional postural displacement (in Russian). *Zh Vyssh Nerv Deiat* 34: 669-677.

Shumilina, A.I. (1949) On participation of pyramidal and extrapyramidal systems in motor activity of a deafferented limb. In Anokhin, P.K. (Ed.), *Problems of higher nervous activity* (in Russian), pp. 176-185. Moscow: AMN SSSR.

Tarnecki, R. (1962) The formation of instrumental conditioned reflexes by direct stimulation of sensori-motor cortex in cats. *Acta Biol Exp* 22: 35-45.

Thomas, E. (1971) Role of postural adjustments in conditioning of dogs with electrical stimulation of the motor cortex as the unconditioned stimulus. *J Comp Physiological Psychol* 76: 187-198.

van Mier, H., Tempel, L.W., Perlmutter, J.S., Raichle, M.E., and Petersen, S.E. (1998) Changes in brain activity during motor learning measured with PET: Effects of hand of performance and practice. *J Neurophysiol* 80: 2177-2199.

Viallet, F., Massion, J., Massarino, R., and Khalil, R. (1987) Performance of a bimanual load-lifting task by parkinsonian patients. *J Neurol Neurosurg Psychiatry* 50: 1274-1283.

Viallet, F., Massion, J., Massarino, R., and Khalil, R. (1992) Coordination between posture and movement in a bimanual load lifting task: Putative role of a medial frontal region including the supplementary motor area. *Exp Brain Res* 88: 674-684.

Wagner, A.R., Thomas, E., and Norton, T. (1967) Conditioning with electrical stimulation of motor cortex; evidence of a possible source of motivation. *J Comp Physiological Psychol* 64: 191-199.

Wentworth, K.L. (1942) Some factors determining handedness in the white rat. *Genet Psychol Monogr* 26: 55-117.

Zelyony, G., Vyssotsky, N., Dobrotina, G., Irzhanskaya, K., Medyakov, F., Naumov, S., Poltyrev, S., and Tuntsova, E. (1937) Kinds and ways of elaboration of associative reflexes. *Proc YI All-Union Congress of Physiol Biochem Pharmacol* (in Russian), pp. 165-171. Tbilisi: USSR Physiological Society.

Zhukov, N.A. (1895) On influence of the motor cortical center lesions on excitability of the adjacent cortical areas. *Doctoral thesis.* SPB University, St. Petersburg.

8

Fractional Power Damping
Model of Joint Motion

James C. Houk
Department of Physiology,
Northwestern University School of Medicine

Andrew H. Fagg and Andrew G. Barto
Department of Computer Science,
University of Massachusetts

No one doubts that the spinal cord plays an important role in movement control. However, when it comes to deciding how motor functions are divided among the brain, the spinal cord, and the musculoskeletal system, one elicits a wide range of opinion. This range of opinion is partly due to the variety of movements that may be considered. For example, basic aspects of locomotion and scratching are built into the spinal circuitry and require the least guidance from supraspinal networks. Voluntary reaching, grasping, and manipulation of objects, in contrast, require extensive guidance from the brain but probably utilize some of the spinal mechanisms that evolved earlier to control locomotion and scratching (Georgopoulos and Grillner, 1989). Postural control is an example of an intermediate behavior, relying heavily both on intrinsic spinal mechanisms and on descending control (Peterson et al., 1992).

This chapter deals specifically with voluntary arm movements. Implications of realistic arm, muscle, and spinal reflex properties for the control of double-joint arms have been investigated by many researchers (Feldman et al., 1990; Flanagan et al., 1993; Gribble et al., 1998; Jordan et al., 1994; Karniel and Inbar, 1997; Katayama and Kawato, 1993; Lukashin et al., 1996; Massone and Myers, 1996; van Dijk, 1978; van Sonderen and Denier van der Gon, 1990). A wide variety of arm-muscle-reflex models are used in these studies since there is no consensus as to what constitutes an adequate model or an appropriate level of abstraction (Winters and Stark, 1987). Although these movements depend heavily on descending control signals, some of their important properties are attributable to spinal reflexes and to musculoskeletal mechanics (Houk and Rymer, 1981).

Our goal is to propose a mathematical model that captures the key complexities of the biological system while preserving a framework that is sufficiently abstract to facilitate computational studies of the overall control problem. One item of complexity that has been neglected in most past studies is the striking dependence of the stretch reflex on a low fractional power of velocity (Gielen and Houk, 1984; Houk, 1981). This nonlinearity produces a frictionlike property in the stretch reflex that probably has a marked influence on voluntary movement control (Barto et al., 1999). The model also incorporates a Hill-based lumped-parameter characterization of the muscle mechanics (Winters, 1990) and many features of the equilibrium-point theory of movement and arm geometry (Bizzi et al., 1992; Feldman et al., 1990; Gribble et al., 1998; Hogan et al., 1987; Mussa-Ivaldi, 1992).

The model architecture is outlined in figure 8.1. The arm model is composed of two joints (a simplified shoulder and an elbow) and moves in the horizontal plane. The arm is actuated by a set of six muscles; a simple model of muscular geometry produces, as a function of skeletal configuration, variations in the muscle's ability to produce torques about the joints. The current muscle length and stretch velocity provide significant influence (through the spinal reflex circuitry) on the muscle's ability to produce forces. A simple pulse-step generator is responsible for production of the descending motor commands. This feedforward control mechanism serves to illustrate the behavior and capabilities of the spinal-musculoskeletal system, but it is not intended as a complete theory of voluntary motor control. In particular, we examine the stiction property of opposing muscles and focus on their participatory role in the production of braking antagonist pulses.

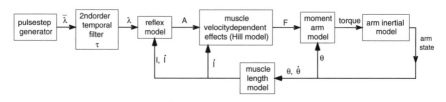

Figure 8.1   Model architecture overview. $\bar{\lambda}$ is the descending muscle motor command, λ is the tonic stretch reflex threshold, A is the motoneuron activation, F is muscle force, θ and $\dot{\theta}$ are joint position and velocity, and l and $\dot{l}$ are muscle length and stretch velocity, respectively.

Musculoskeletal Geometry

Our skeletal model represents a human arm operating in a horizontal plane with two degrees of freedom: rotation of the shoulder and elbow (see figure 8.2). We use the standard equations of motion (e.g., Hollerbach and Flash, 1982) with the following parameters after Gribble et al. (1998)—mass of upper and lower arm: 2.1 kg (4.6 lb) and 1.65 kg (3.6 lb); length of upper and lower arm: 34 cm (13.4 in.) and 46 cm (18.1 in.); moment of inertia about center of mass of upper and lower arm: 0.023 kg/m^2 (0.549 lb/ft^2) and 0.011 kg/m^2 (0.258 lb/ft^2). Unlike some other models, the lower arm includes the hand, although the wrist is assumed to be locked, which is appropriate for our studies of arm motion. We lump the set of muscles acting on the arm into three pairs of equivalent muscles (e.g., Winters and Stark, 1988). One pair consists of a flexor and an extensor representing all the synergistic one-joint muscles for the shoulder, the actions of which are assumed to be dominated by the pectoralis and deltoid, respectively. A second pair of one-joint muscles model those acting on the elbow, which are assumed to correspond roughly to the brachialis and the triceps lateral head. The third pair represents flexor and extensor biarticulate muscles spanning both joints, corresponding to the actions of the biceps and the triceps long head, respectively. The muscle moment arms for the extensors are set to 3.5 cm (shoulder extensor; 1.4 in.), 2.5 cm (elbow; 1 in.), 4 cm (biarticulate shoulder; 1.6 in.), and 2 cm (biarticulate elbow; 0.8 in.). The flexor moment arms are assumed to vary between 0 and 5 cm (0 and 2 in.), depending on the configuration of the arm.

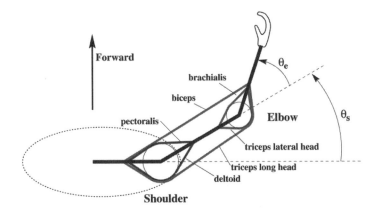

Figure 8.2   A schematic view of an idealized arm model with six muscles. $\theta_e = 0°$ (elbow orientation of 0°) is defined as full extension; $\theta_s = 0°$ is defined as the upper arm aligned with the sagittal plane of the body. The biceps and triceps long head produce moments about both the shoulder and elbow. The extensors are assumed to wrap around spherical joint capsules throughout the range of motion (resulting in constant moment arms). For the flexors, the muscles are assumed to leave the joint capsule at a critical flexion threshold, beyond which the muscles are modeled as following a straight path from origin to insertion. The origins and insertions depicted are not to scale.

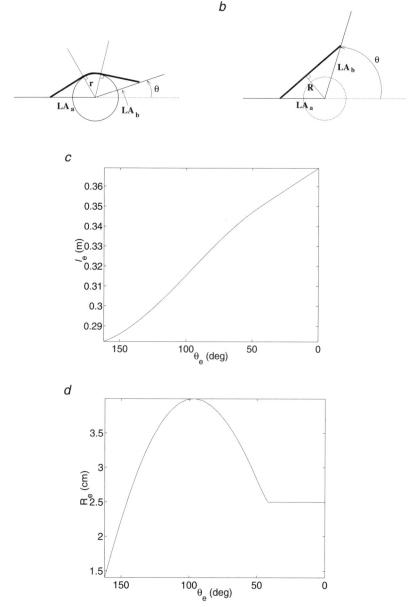

Figure 8.3 A simple muscle path model: *(a)* wrapped path and *(b)* straight path. Muscle path follows the dark curves. The critical parameters/variables are θ (joint orientation, $\theta_e = 0° =$ full elbow extension), LA_a and LA_b (distance from center of joint rotation to muscle origin and insertion, respectively), r (radius of the joint capsule), and R (muscle moment arm). This model has critical implications for both *(c)* the muscle length and *(d)* the muscle moment arm as the skeletal configuration changes.

Winters and Stark (1988) suggested a simple model of muscle path in which one assumes that, for extended joint positions, the muscle wraps around a spherical joint capsule, resulting in a constant moment arm (figure 8.3a). However, when the joint flexes beyond a critical threshold, we assume that the muscle leaves the joint capsule and follows a straight path from origin to insertion (figure 8.3b). The result is a muscle moment arm that can be larger than the joint capsule, as demonstrated in figure 8.3d. Note that for extreme flexion (where $LA_a \neq LA_b$), the moment arm can also drop to a level below that of the joint capsule radius.

The work of Amis et al. (1979) and An et al. (1981) indicates that this form of path model captures the primary variation of muscle moment arms as a function of joint orientation for a number of elbow muscles, including the brachialis. Less is known about the geometry of muscles involving the shoulder, and we assume for simplicity that this path model also applies in this case.

Little is also known about the geometry of the muscles that actuate two joints. We assume a generalization of the model mentioned earlier. In this case, the relationship between muscle path and joint configuration is more complicated, as the muscle may wrap around either one, both, or neither of the joint capsules. The two extreme cases are shown in figure 8.4. The muscle moment arms as a function of joint configuration are demonstrated in figure 8.4, c and d. Further details of derivation

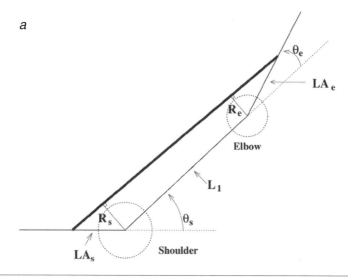

Figure 8.4 Two cases of the biarticulate muscle path model: *(a)* no contact with the joint capsules and *(b)* contact with both joint capsules. Biarticulate moment arm as a function of joint angle for the *(c)* shoulder and *(d)* elbow. θ_s and θ_e correspond to the joint angles for the shoulder and elbow, respectively; L_1, LA_s, and LA_e represent the distance from shoulder to elbow joint, the distance from muscle origin to center of shoulder rotation, and the distance from muscle insertion to the center of elbow rotation; r_s and r_e are the joint capsule radii; and R_s and R_e are the moment arms for the shoulder and elbow. For the case illustrated in *(b)*, $R_s = r_s$ and $R_e = r_e$.

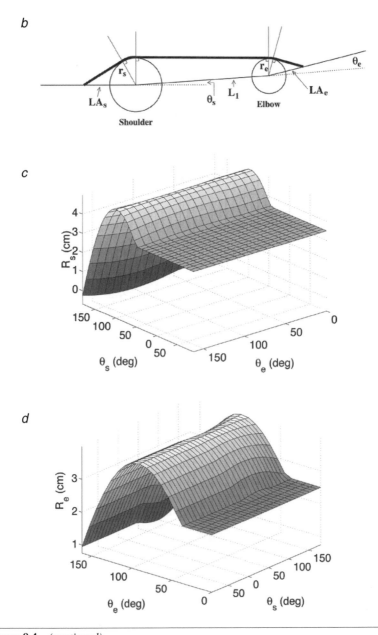

Figure 8.4   *(continued)*

of the geometric muscle model, as well as the assumptions about the critical parameters (including joint capsule radii and muscle origin/insertion locations) may be found in Fagg (2000).

The Tonic Stretch Reflex

As in the λ model of Feldman and colleagues (Feldman, 1966; Feldman et al., 1990), the neural control signal in our model determines the threshold muscle length, λ, for initiation of the tonic stretch reflex. Static force (zero stretch velocity), F, is generated as a function of the difference between a muscle's current length, l, and the current value of its λ, as plotted in figure 8.5a. Specifically:

$$F = K[l - \lambda]^+ \left(1 - e^{-\frac{[l-\lambda]^+}{c}} \right),$$
(8.1)

where

$$[x]^+ = \begin{cases} x & \text{if } x > 0; \\ 0 & \text{if } x \le 0. \end{cases}$$

The exponential term captures the initial recruitment of the motor units in a fashion that is related to the size principle (Binder et al., 1996; Houk et al., 1970). This term influences the force-length relationship most strongly at lengths just above the stretch reflex threshold, λ. The parameter c determines the spatial extent of this influence. A short derivation is given in the appendix, and the values used for the individual muscles are shown in table 8.1. As $[l - \lambda]^+$ increases, force due to the tonic stretch reflex approaches a linear function of length, with slope K (figure 8.5a). This parameter represents the stiffness of the tonic stretch reflex.

The *normalized stiffness* of a muscle is defined as the stiffness of the stretch reflex normalized by the operating range of the muscle or joint. Houk and Rymer (1981) observed that across different muscles, normalized stiffness (K_n) tends to take on a constant value of about unity. This provides a convenient method by which reflex stiffness values may be computed from estimates of the operating range of muscle length and muscle force:

$$K = K_n \frac{\text{force range}}{\text{length range}},$$
(8.2)

where *length range* is the change in muscle length from full extension to full flexion. We assume for our purposes that $K_n = 1$. Minimum force is assumed to be 0; thus, *force range* is taken to be *max force,* which is often assumed to be linearly related to the muscle's physiological cross-sectional area (PCSA) (An et al., 1981). The derivation of this latter transformation is given in the appendix. The muscle parameters and resulting reflex stiffnesses are shown in table 8.1.

The force-length behavior for a muscle is illustrated in figure 8.5a (muscle stretch velocity is 0). Changes in the descending motor command, λ, result in a shift of the force-length curve along the length axis (to the left for decreasing values of λ, to the right for increasing values), as in the λ model (Feldman, 1966). The combined effect of all muscles acting on a single joint is derived by appropriately summing the corresponding force-length curves (e.g., Partridge, 1979), weighted by the muscle moment arm:

$$\tau_j = \sum_{m \in M_j} R_{j,m}(\theta) F_m(\theta, \lambda), \tag{8.3}$$

where τ_j is the torque exerted at joint j, M_j is the set of muscles that actuate joint j, $R_{j,m}(\theta)$ is the moment arm of muscle m about joint j, F_m is the force produced by the tonic stretch reflex for muscle m, and θ is the two-element vector composed of the shoulder and elbow flexion angles. Note that we utilize the convention that $R_{j,m}(\theta)$ is positive if the muscle generates joint flexion and negative if it produces joint extension. We assume that the descending motor command affects both of the reflex

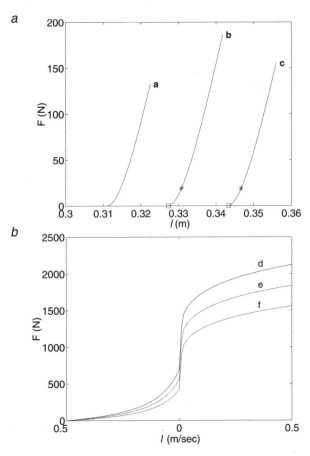

Figure 8.5 (a) Force as a function of muscle length for the simulated brachialis. It is assumed that the muscle is not stretching (i.e., $\dot{l} = 0$). (a) $\lambda = 31.16$ cm (12.27 in.); (b) $\lambda = 32.85$ cm (12.93 in.); (c) $\lambda = 34.46$ cm (13.57 in.). The four marked points (two stars and two boxes) correspond to the same configuration and motor command as those indicated in figure 8.6a. (b) Force as a function of muscle stretch velocity ($\dot{l}$) for the simulated brachialis. $\dot{l} < 0$ corresponds to shortening of the muscle. $\lambda = 33.6$ cm (13.23 in.). (d) $l = 36.28$ cm (14.28 in.); (e) $l = 35.49$ cm (13.97 in.); (f) $l = 34.68$ cm (13.65 in.).

Table 8.1   Muscle Parameters

Muscle	PCSA (cm^2)	Moment arm (cm)	Length range (cm)	K (N/m)	H $\left(\frac{\sec}{m}\right)^{\frac{1}{5}}$	c (cm)
Pectoralis	6.8 Freivalds (1985)	3.5 – (3.5) – 5	15.29	9,022	0.61	1.56
Deltoid	11.01 Freivalds (1985)	3.5	13.74	16,251	1.10	1.41
Brachialis	7.0 An et al. (1981)	1.4 – (2.5) – 4	8.91	15,942	1.08	0.91
Triceps lateral head	6.0 An et al. (1981)	2.5	7.85	15,498	1.05	0.80
Biceps	4.6 (short head + long head) An et al. (1981)	0 – (3.5) – 5 (s) 0.75 – (2.5) – 4 (e)	22.86	4,083	0.28	2.34
Triceps long head	6.7 An et al. (1981)	4 (s) 2.5 (e)	23.56	5,769	0.39	2.41
Flexor carpi radialis	2 An et al. (1981)	2	4.71	8,610	0.59	–

Notation for variable moment arms: "minimum – (constant) – maximum," where *constant* refers to the constant moment arm size in the wrapping region. The parameters of the flexor carpi radialis are used in the appendix to derive a number of other model parameters. PCSA = physiological cross-sectional area; K = spring constant; H = nonlinear damping constant; c = stretch reflex onset parameter; s = shoulder; e = elbow.

thresholds ($\lambda_{agonist}$ and $\lambda_{antagonist}$) by simultaneously increasing one as the other is decreased (or vice versa). We also include a descending co-contraction signal (described in the section "Responses to Voluntary Commands"), which is implemented as a simultaneous decrease in the two reflex thresholds. When $\lambda_{agonist}$ and $\lambda_{antagonist}$ are selected in such a way that both muscles are tonically active, equation (8.3) defines a force field with a unique equilibrium position (Bizzi et al., 1992; Hogan et al., 1987).

The combined effect of the four equivalent muscles acting on the elbow joint is shown in figure 8.6a. The curves illustrated in the figure demonstrate the position-dependent, isometric torque response of the set of muscles under three different constant motor commands. The motor commands were selected under the same conditions as those produced by subjects of the Astryan and Feldman (1965) experiment (the results of which are shown in figure 8.6b). A specified level of torque is established against a load at a specified elbow position (each indicated by one of the three circles in figure 8.6a). This is accomplished with the model by setting the elbow to the desired position and then shifting λ's together until the desired torque is achieved. Redundant degrees of freedom are resolved by constraining the equilibrium position of each agonist/antagonist pair (brachialis/triceps lateral head and biceps/triceps long head) to be the same joint position and by enforcing a small amount of overlap in the tonic reflex region for each pair (i.e., each pair is slightly co-contracted).

The curve emanating from each of the circles is produced by holding the motor commands constant as the elbow is allowed to flex. The resulting muscle-produced torque is thus equal in magnitude but opposite in direction to the opposing isometric load at that joint location. In other words, if the opposing load were suddenly reduced from its initial level (at the circles) to some lower magnitude, the curves indicate the position to which the elbow would equilibrate (assuming no changes in the descending motor commands). The precise shape of the isomotor command curves is determined by a combination of the force-generation ability of the tonic

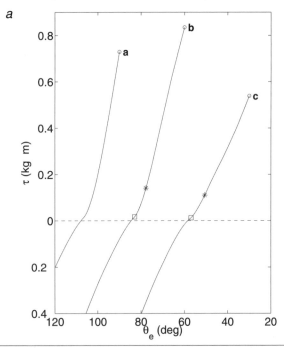

Figure 8.6   *(a)* Torque as a function of elbow position (θ_e) for three levels of constant motor command. Elbow position is in the coordinate system of our model, with $\theta_e = 0$ corresponding to full elbow extension. *(b)* Observed torque as a function of elbow position (reprinted from Astryan and Feldman, 1965). Note that $\psi = 180 - \theta_e$. *(c)* Torque as a function of elbow position (θ_e). Data taken under the same conditions as in figure 8.6a, except that a wider region of state space and descending motor commands is shown. The dashed box indicates the same region of space as in figure 8.6a.

Figure 8.6b adapted from Asatryan and Fel'dman 1965.

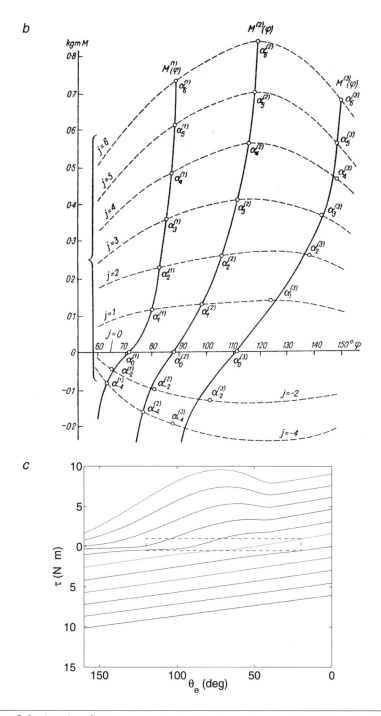

Figure 8.6 *(continued)*

stretch reflex and the position-dependent variation in muscle moment arm. The difference in slope in the curves in figure 8.6, b and c results from the increase in moment arm due to a lifting of the muscle off of the joint capsule. The contribution of the nonlinearity of the stretch reflex versus the moment arm variation becomes clearer by comparing the corresponding points in figures 8.5a and 8.6a (the points are marked with stars or boxes in the two figures). Furthermore, figure 8.6c shows the elbow torque response over a wider range of the variables. The change in slope of the isomotor command curves is a direct result of the variation in muscle moment arms for the elbow flexors. The constant slope of these curves in the $\tau < 0$ region is due to our assumption of constant moment arms for extensors. This assumption roughly approximates the experimental observations of Amis et al. (1979) regarding the elbow joint and is consistent with the assumptions made by the models of Asatryan and Feldman (1965) and Gribble et al. (1998).

Fractional Power Damping

Our model of muscle and spinal reflex properties also includes significant velocity-dependent components. As in the λ model and the muscle-reflex model of Wu et al. (1990), the force-velocity-length relation includes the effects of reflex-induced electromyographic (EMG) activity, which particularly influences the relation for positive velocities (lengthening). For negative velocities (shortening), we ignore stretch reflex velocity dependence where it has considerably less effect on force generation (Gielen and Houk, 1984). Instead, the variation of force production with shortening velocity is assumed to be dominated by muscle mechanical properties. In contrast, the velocity dependence of lengthening responses is modeled as the combined result of muscle mechanics and the stretch reflex (Gielen and Houk, 1984; Houk, 1981). We also adopt the well-supported approach of multiplicatively combining length- and velocity-dependent muscle mechanical characteristics (Winters, 1990; Wu et al., 1990).

Combining the approach of Gribble et al. (1998) and Wu et al. (1990), we define muscle activation at time t to be

$$A = [l - \lambda + Hr]^+.$$ (8.4)

The terms $l - \lambda$ and Hr, respectively, represent the static and dynamic components of reflex-produced activation, where H is a gain coefficient for the dynamic component. The values used for specific equivalent muscles are given in table 8.1. The reflex-related damping term, r, is defined as follows:

$$r = \begin{cases} \left[\dot{l}^{1/5}(l - \lambda + \mu) \right]^+ & \text{if } V < \dot{l} \text{ (fast lengthening);} \\ \left[V^{1/5}\dfrac{\dot{l}}{V}(l - \lambda + \mu) \right]^+ & \text{if } 0 \leq \dot{l} < V \text{ (slow lengthening);} \\ 0 & \text{otherwise (shortening),} \end{cases}$$ (8.5)

where $V = 1.25$ cm/s (0.49 in./s) is a stretch velocity threshold separating the "fast" and "slow" lengthening regions and $\mu > 0$ represents the receptor's baseline positional sensitivity (described later).

The rationale for this definition of the dependence of muscle activation on stretch velocity follows Wu et al. (1990). In ramp stretch experiments with the human wrist, Gielen and Houk (1987) estimated power law relationships for EMG activity and force response to have exponents of approximately 1/3 and 1/5, respectively. Modeling studies (Gielen and Houk, 1984; Houk, 1981) suggest that the smaller exponent for force results from the combination of muscle mechanical properties and reflex-produced neural input during muscle lengthening. For mathematical convenience in the fractional power damping (FPD) model, we use the 1/5 exponent to define the velocity-dependent muscle activation during lengthening. However, the slope of $\dot{l}^{1/5}$ approaches infinity as $\dot{l}$ approaches 0, which results in numerical instabilities in the model. This difficulty is solved by introducing a short, linearly varying region in the range $0 \leq l < V$ (corresponding to the "slow lengthening" component of equation [8.5]). Following the results of Houk et al. (1970), which indicate a small contribution of tendon organ feedback to motoneuron activity, a tendon organ contribution to muscle activity is not explicitly included in the model (cf. Houk and Rymer, 1981).

Setting the parameter μ in equation (8.5) to be greater than zero allows reflex activity when the muscle length is below the threshold λ of the tonic stretch reflex, provided that the muscle is lengthening. It therefore has an effect similar to that of parameters μ of Gribble et al. (1998) and x_{p0} of Wu et al. (1990). We choose $\mu = 0.028$ m (0.092 ft), which falls within the upper range estimated by Wu et al. (1990).

In the dynamic case, equation (8.1) is expanded to include velocity-dependent muscle activation and muscle mechanical effects. Specifically,

$$F = K A \left(1 - e^{-A/c}\right) m\left(\dot{l}\right), \tag{8.6}$$

where

$$m\left(\dot{l}\right) = \begin{cases} \left[\left(b + a\dot{l}\right) \middle/ \left(b - \dot{l}\right)\right]^{+} & \text{if } \dot{l} \leq 0 \text{ (shortening);} \\ 1 & \text{otherwise (lengthening),} \end{cases} \tag{8.7}$$

and where a and b are the Hill equation parameters (Hill, 1938). K and c are the reflex stiffness and reflex threshold transition parameters, respectively (as defined for equation [8.1]). Although $m(\dot{l}) = 1$ appears to imply that there is no drop in force production with a positive stretch velocity, these effects are actually accounted for by choice of the 1/5 fractional power of equation (8.5).

For muscle shortening, equation (8.7) captures the Hill equation in which the maximum shortening velocity, v_{max}, is $-b/a$. The FPD model incorporates the simplest assumption that v_{max} remains constant over muscle length and activation level. Winters (1990) discusses implications of this assumption, which we consider to be minor for our purposes. Also for simplicity, we assume that a and b are the same for all the model's muscles, although these parameters would vary as a function of muscle fiber length in a more detailed model. We set $a = 0.25$ following Winters (1990). The value for b is determined by averaging the values for v_{max} given by Winters and Stark (1985) in radians per second. Assuming a constant moment arm of 2.0 cm (0.79 in.) yields $v_{max} = -0.5$ m/sec (–1.64 ft/sec), or $b = 0.125$.

We assume that the nonlinear damping gain, H, is proportional to a muscle's stiffness, K. Gielen and Houk (1984) reported a force relationship to velocity for stretching of the wrist flexor muscles described by the following equation:

$$F - F_0 = C_1 v^{1/5}(x - x_{01}),$$

where F is the force produced against the manipulandum, F_0 is the force prior to stretch, v is the velocity of manipulandum movement, x is the current manipulandum position, and x_{01} represents the combined effects of λ and μ in equation (8.5). In the FPD model, KH corresponds to C_1, for which Gielen and Houk (1984) estimated a value of approximately

$$955 \frac{N}{m} \left(\frac{sec}{m} \right)^{1/5}$$

In this experiment, length (x) was measured in the coordinate system of the manipulandum, which had a moment arm of approximately four times that of the wrist-actuating muscles. Thus, for our purposes:

$$C_1 = 955 \times 4 \times 4^{1/5} = 5,040.5 \frac{N}{m} \left(\frac{sec}{m} \right)^{1/5}.$$

If we assume a stiffness of $K = 12,069$ N/m for the muscles involved in wrist flexion (a quantity that is derived in the appendix), then H_{fcr}, the lumped gain parameter for the wrist flexion muscles,[1] is defined as follows:

$$
\begin{aligned}
H_{fcr} &= \frac{C_1}{K} \\
&= 5040.5 \frac{N}{m} \left(\frac{sec}{m} \right)^{1/5} \frac{1m}{8,610 \, N} \\
&= 0.59 \left(\frac{sec}{m} \right)^{1/5}.
\end{aligned}
\tag{8.8}
$$

The damping gain parameter for any particular muscle (H_i) is related to the muscle's stiffness relative to the flexor carpi radialis:

---

[1]We assume an equivalent muscle corresponding to the flexor carpi radialis.

$$H_i = H_{fcr} \frac{K_i}{K_{fcr}}.$$

The resulting parameters that were used in the model are given in table 8.1.

As described previously, experimental results suggest that EMG activation varies as a 1/3 fractional power of stretch velocity (Gielen and Houk, 1984; Houk, 1981). When relating model behavior to EMG data, we utilize a modification of equation (8.4) in which the 1/5 fractional power (of equation [8.5]) is replaced with 1/3 power. Note, however, that this formation omits the initial burst-phase of spindle responses to stretch (Hasan and Houk, 1975; Houk et al., 1992). Specifically:

$$EMG = \begin{cases} \left[l - \lambda + H \left[\dot{i}^{1/3} \left(l - \lambda + \mu \right) \right]^+ \right]^+ & \text{if } V \upharpoonleft \dot{i}; \\[2ex] \left[l - \lambda + H \left[V^{1/3} \frac{\dot{i}}{V} \left(l - \lambda + \mu \right) \right]^+ \right]^+ & \text{if } 0 \leq \dot{i} < V; \\[2ex] \left[l - \lambda \right]^+ & \text{otherwise.} \end{cases} \quad (8.9)$$

The velocity-dependent effects of the Hill equation for muscle shortening and the Gielen-Houk stretch reflex nonlinearity are illustrated in figure 8.5b. The three curves represent different constant muscle lengths. As shortening velocity approaches V_{max} (–0.5 m/sec [– 1.64 ft/sec]), force generation ability drops to 0 N in all cases. With muscle stretch, there is an initial rapid increase in the response force. The slope of this response drops as stretch velocity increases.

The combined effect of static and dynamic components is illustrated in figure 8.7. In this case, $\lambda = 0.323$ m (1.06 ft); changes in λ shift the surface along the length axis. An important feature of this surface is that for any fixed stretch velocity, force (F of equation [8.6]) approaches a linear function of l whose slope increases with increasing l. This results from the $l - \lambda$ term that appears in both the definition of muscle activation, A (equation [8.4]), and the definition of spindle activation, r (equation [8.5]). Note that although the reflex threshold is set to 0.323 m (1.06 ft), during stretch, activation onset occurs much earlier (about 0.315 m [1.03 ft] in figure 8.7). This effect is due to the additive combination of the position- and velocity-sensitive muscle spindles in equation (8.4).

FPD Interaction of Multiple Muscles

A critical feature of the force-length-velocity curve of figure 8.7 is the sudden increase in response force as the system transitions from zero stretch velocity to a small degree of stretching. At the initiation of a muscle stretch, this immediate response is due to the large short-range mechanical elasticity of the muscle, which is then followed by a delayed reflex response (Houk, 1981; Houk et al., 1981).

However, the characterization captured by equations (8.4) and (8.5) lumps the two components together, allowing a simpler mathematical specification.

When the reflex thresholds for an agonist/antagonist muscle pair are set such that both muscles are simultaneously active, then the result is a region of *stiction* that surrounds the equilibrium position in the force field produced by the muscle pair. This stiction region is characterized by a rapid change in the torque response with either a positive or negative deviation from zero stretch velocity. This is demonstrated in figure 8.8, which shows the torque response as a function of elbow position and velocity, assuming a constant set of descending motor commands. The reflex thresholds are set such that the joint equilibrium position is 90°, which corresponds to the center of the ellipse. The region corresponding to low joint position values and negative joint velocities is the stretch region for the flexors acting on the elbow, which produce a large positive torque. The bold S-shaped curve indicates where torque $\tau = 0$. The hashed line is where $\dot{\theta} = 0$ (the tonic torque reflex response).

Joint positions near the equilibrium are such that small negative joint velocities (joint extension/stretch of the elbow flexors) result in a large opposing response torque (ellipse region of figure 8.8). Positive joint velocities also result in a large opposing response torque. When the joint state approaches the equilibrium (within about ±20° from the target at 90°), this combination of opposing forces has the effect of very quickly driving the joint velocity to a very small level. When this happens, although the joint continues to move very slowly toward the equilibrium, behaviorally, the joint effectively stops moving (or "sticks") at the point at which it entered the stiction region. Once within this region, the joint requires a significant pertur-

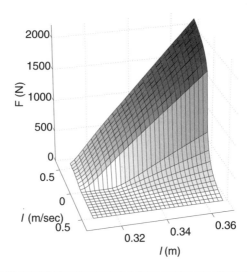

Figure 8.7   Force as a function of muscle length (l) and velocity ($\dot{l}$) for both shortening and lengthening with $\lambda = 0.323$ m (1.06 ft). Positive velocity corresponds to muscle stretching.

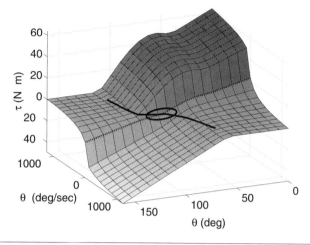

Figure 8.8   Elbow torque as a function of configuration of the elbow (θ) and its velocity ($\dot{\theta}$). The action of four muscles contributes to the total torque. $\theta = 0$ corresponds to full extension of the elbow. The shoulder is held at a fixed position. The bold S-shaped curve indicates where torque $\tau = 0$; the ellipse indicates the stiction region. The equilibrium position of the joint is located in the middle of the stiction region (90°).

bation or shift in equilibrium position to be dislodged. The result is a system that does not exhibit significant ringing near the equilibrium (Barto et al., 1999).

Far away from the equilibrium position, this stiction property does not hold. Although there is a significant response torque for deviations from zero velocity in one direction, an opposite response does not exist for deviations in the other direction. In fact, the response torque does not change sign, as it does within the stiction region.

Figure 8.9 shows the phase plane dynamics corresponding to the torque surface shown in figure 8.8. Each arrow indicates the direction and magnitude of the evolution of the system given the state corresponding to the tail of the arrow. The bold curve is the velocity nullcline ($\dot{\theta} = 0$). Note that this set of points differs from that of figure 8.8, in which $\tau = 0$; the velocity nullcline includes forces in addition to those produced by the muscles (including inertial, coriolis, and externally induced forces). The upper right and lower left quadrants of figure 8.9 correspond to the regions of state space in which the flexor and extensor muscles (respectively) respond to stretch by producing large forces. As seen in the figure, these forces result in a rapid reduction in velocity.

In the model, one class of motor commands (described later) causes the equilibrium positions of muscle pairs to shift together. This has the effect of shifting the entire surface of figure 8.8 along the joint position dimension. Co-contraction of muscle pairs has the effect of bringing the two stretch reflex regions closer together along the position dimension, thus increasing their overlap. This results in an increase in the width of the stiction region.

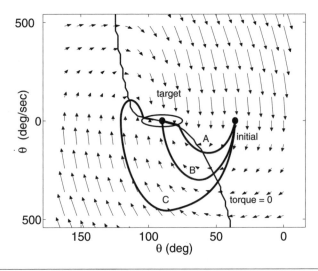

Figure 8.9 Phase plane dynamics for the elbow joint. The arrows indicate the direction and magnitude of the evolution of the elbow state (θ and $\dot{\theta}$). The bold S-shaped curve represents the velocity nullcline ($\ddot{\theta}=0$). The muscle motor commands are such that the equilibrium point is at 90°, which falls in the center of the ellipse indicating the stiction region. Three paths (emanating from the *initial* position at 36°) show the evolution of elbow state for three different movement trials, in which different motor commands are utilized (these are described in the section "Responses to Voluntary Commands"). Path A corresponds to the case in which the equilibrium position is shifted to the target without first executing a pulse. In this case, the elbow effectively falls short of the target. Path B is one in which the pulse magnitude and duration are such that the elbow stops at the target. Path C is one in which the pulse magnitude is too large, resulting in an overshoot of the target.

Responses to Voluntary Commands

Arm movements are controlled in the model by producing time-varying motor commands (λ_m's) that descend to the spinal cord. One of the simplest representations of such a time-varying behavior is the pulse-step motor command. This form of motor command description is well established in the saccadic eye movement literature (Robinson, 1975) and has been suggested as a reasonable approximation for the control of voluntary limb movements (Ghez, 1979; Ghez and Martin, 1982). The pulse-step waveform describes the time course of the reflex thresholds for an opposing pair of muscles. Hence, we can think of the waveform as specifying the equilibrium position of the joint, with the pulse playing the role of movement initiation and the step indicating where the movement should stop. However, unlike other equilibrium point theories, simply setting the step to the target joint location in the FPD model does not guarantee that the joint will arrive there in a reasonable amount of time (Barto et al., 1999). Instead, due to the stiction property around the equilibrium, the joint may effectively stop at a point that is not the equilibrium. As a result, it is necessary to specify the properties of the pulse (height and duration) such that the joint sticks at the desired target.

In addition to specifying the time course of the equilibrium position, it is also possible to superimpose a co-contraction signal that causes an overlap in the tonic muscle response of the opposing muscle pair. For simplicity, here we will assume that a constant nonzero level of co-contraction is specified prior to and during the entire movement.

For the case of a single opposing pair of muscles, the individual reflex thresholds ($\bar{\lambda}_{flexor}$ and $\bar{\lambda}_{extensor}$) are expressed as follows:

$$\bar{\lambda}_m(t) = CC + \begin{cases} L_m(\theta_I) & \text{if } t < 0; \\ L_m(\theta_T) + D_m P & \text{if } 0 <= t < S; \\ L_m(\theta_T) & \text{if } t <= S, \end{cases} \tag{8.11}$$

where CC is the co-contraction level used throughout this example; $L_m(\theta)$ is the commanded length of muscle m at the corresponding elbow joint angle, θ; θ_I and θ_T are the equilibrium positions for the initial and target positions; D_m is the direction of muscle pull (+1 for flexor and –1 for extensor); P is the pulse magnitude; and S is the time of transition from pulse to step.

The muscle pulse-step waveforms ($\bar{\lambda}_m$'s) are made less abrupt by temporally filtering them through a cascade of two low-pass filters, each with a time constant of 40 ms (Engelbrecht, 2001). Miller and Sinkjaer (1998) suggest that such a form of temporal filtering may be the result of the time required to recruit a large set of cortical motoneurons and their associated interneurons. The result is a set of motor signals (λ_m's) representing the spinal reflex thresholds of equations (8.2)-(8.7). Note that this form of motor program does not allow for the online adjustment of the descending motor signals. Instead, the parameters are selected a priori and the motor commands are executed in a completely open-loop manner. Thus, our use of a pulse-step waveform should not be viewed as a complete theory of limb motor control. Instead, this mechanism should be seen as a technique for "exercising" the plant model, which incorporates skeletal, muscular, and spinal mechanisms.

For the results that follow, the experimental paradigm is one in which the shoulder is held in a fixed position ($\theta_s = 90°$) and the model is asked to generate point-to-point movements of the elbow. Because both mono- and biarticulate muscles are involved in production of elbow movement, we are faced with an additional level of redundancy. For simplicity, CC, P, and S are chosen to be the same for both pairs of muscles acting on the elbow.

Figure 8.10 demonstrates the behavior that results from the execution of three different pulse-step waveforms. Each of the three movements starts with the elbow at 36°. The target is set to 90°. We selected $CC = 0.015$ for all three pulse-step waveforms, which yields a movement duration of about 300 ms for the accurate reach (figure 8.10b). For each of the three cases, five subpanels are shown. The first two subpanels demonstrate the time course of the elbow joint angle and angular

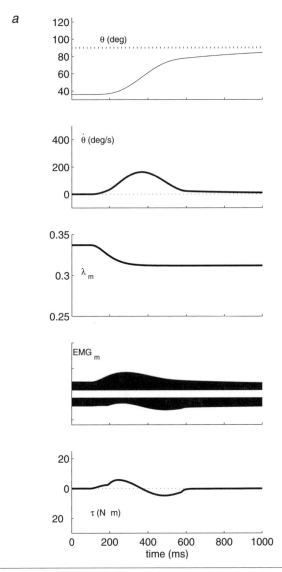

a

Figure 8.10 Behavioral patterns that result from executing three different pulse-step waveforms. *(a)* Execution of a step only (the equilibrium point is shifted directly from initial to target positions. *(b)* Execution of a pulse-step that brings the elbow directly to the target position. *(c)* Execution of a pulse-step that results in an overshoot of the target. The vertical subpanels show the time courses of the following: elbow position, elbow velocity, muscle motor command for the monoarticulate flexor, simulated EMG activity for the brachialis (dark region) and triceps lateral head (light region), and the muscle-produced torque. Magnitude of EMG activity is indicated by the height of the shaded regions. The activity of the brachialis (the flexor) is shown as a deviation in the upward direction; the activity of the triceps lateral head (extensor) is shown as a deviation in the downward direction.

b

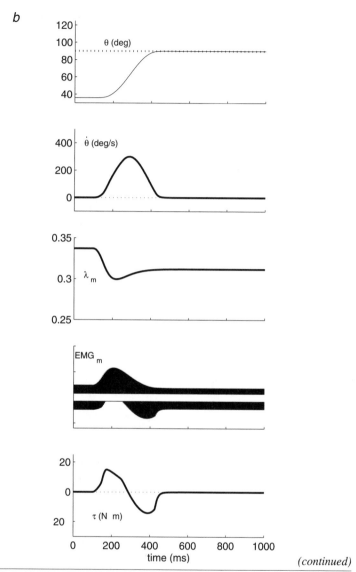

(continued)

Figure 8.10   *(continued)*

velocity, respectively. The third subpanel shows the stretch reflex threshold (λ_m) for the monoarticulate flexor (agonist). The fourth subpanel illustrates the time course of the EMG signals for the monoarticulate flexor (positive deviation) and extensor (negative deviation). EMG magnitude for the corresponding muscle is read as the height of the shaded region. The final subpanel indicates the muscle-induced torques that result from the descending motor commands.

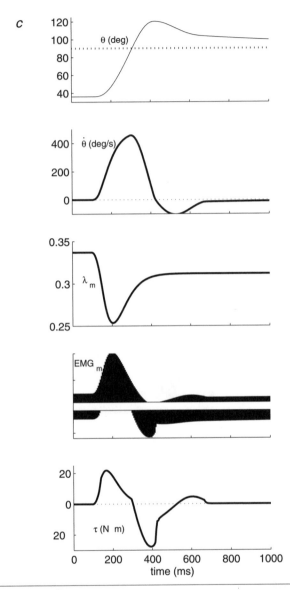

Figure 8.10 *(continued)*

Figure 8.10a shows the case in which no pulse is executed ($S = 0$). The result is a movement that undershoots the target by more than $12°$. Once the elbow movement slows down to a low velocity at about 600 ms, it enters the elbow stiction region. Once there, it maintains a very slow creep toward the equilibrium position. Even after an additional 400 ms, the elbow is still about $5°$ from the target. Thus, for all intents and purposes (certainly on a behavioral time scale), the movement has effectively stopped at 600 ms. The corresponding phase plane behavior is shown in figure 8.9a. The path obtains a low-magnitude peak velocity and enters the stiction region on the right-hand side of the target position.

The descending motor command (middle subpanel of figure 8.10a) shows a decrease in the stretch reflex threshold for the monoarticulate flexor. There is a simultaneous increase in the threshold for the extensor that is not shown. The result is a shift in the equilibrium position from the initial position to the target. The gradual transition from the initial position ($\lambda_m = 33.6$ cm [13.23 in.]) to the target ($\lambda_m = 31.1$ cm [12.24 in.]) is due to the temporal filtering in the pulse-step waveform (see figure 8.1). The reduction in the flexor threshold results in a reflex-induced increase in the EMG activity of the flexor. The opposite effect is seen in the extensor EMG. The combined effect is a net positive torque produced by the muscles, which initiates an elbow flexion movement.

The transition to the braking phase of movement is the result of two distinct components (from equation [8.4]): (1) the movement of the elbow increases the length of the extensor, increasing the tonic response of the stretch reflex, and (2) this response is further facilitated by the dynamic response of the stretch reflex. Thus, prior to reaching the target, the torques induced by the extensor overcome those of the flexor, resulting in a net negative (or slowing) torque. After the movement settles to a low velocity, the residual EMG activity in both the flexor and extensor is primarily the result of the descending co-contraction signal.

When a pulse is introduced with appropriate parameters ($S = 50$ ms and $P = 0.065$), the elbow completes its motion exactly at the target (figure 8.10b and path B of figure 8.9). The direct effects of the pulse are seen in the trace of the muscle motor command (middle subpanel). Compared to the case in which no pulse is executed, these thresholds temporarily achieve a higher deviation from the target step level. This difference leads to an even higher level of EMG activity for the flexor, which results in a higher acceleration of the elbow and hence an increase in peak velocity. Because the elbow obtains a higher velocity than the step-only case, the extensor stretch velocity is higher, leading to an increase in the dynamic response of the antagonist stretch reflex. This results in a larger braking response of the extensor muscles.

In the case in which the pulse magnitude is set even higher ($P = 0.2$), the elbow overshoots the target (figure 8.10c). As seen in the phase portrait (figure 8.9, path C), the path completely passes the stiction region. As a result, the elbow begins to move in the opposite direction. However, this re-acceleration is limited; the resulting stretch of the flexor induces a second agonist burst, which again slows the elbow, causing it to fall into the stiction region (this time, the left-hand side of the target position). Note that in the phase plane, the second velocity peak corresponds (by definition) to the path's crossing of the velocity nullcline ($\ddot{\theta} = 0$). The reason that the first velocity peak does not occur at the pictured nullcline is that at the time of the peak, the descending motor signals do not correspond to the target but instead to the target "plus pulse" (with temporal smoothing). Thus, at the time of the first velocity peak, the nullcline is actually shifted to the left of where it is depicted in figure 8.9.

The torque profile for the overshoot case shown in figure 8.10c (the fifth subpanel) contains two sharp changes in slope at about 300 ms and 430 ms. These points correspond to the onset and offset of the antagonist muscle burst. It is important to note that although the descending motor signals (the λ_m's) specify the time of transition to the step, and the step level itself, the precise onset time and magnitude of the antagonist braking pulse is determined by the current state of the arm (θ and $\dot{\theta}$). Even when the descending command remains the same on a different movement trial, if the phase plane evolution of the arm shows any differences (e.g., due to external loading), this will be reflected as a difference in the antagonist burst pattern.

Conclusion

This chapter focuses on the development of a neuromuscular model that captures a tradeoff between the key complexities of the biological system and the preservation of an abstract framework that facilitates efficient simulation of the model. In particular, it is important to take into account certain properties of the muscle mechanics and the spinal reflex circuitry. To that end, our model represents a synthesis of equilibrium point approaches (e.g., Bizzi et al., 1982; Feldman, 1966; Feldman et al., 1990; Gribble et al., 1998), theories of velocity-dependent force generation of muscle (Winters, 1990; Winters and Stark, 1988), and theories of stretch-reflex-induced nonlinear damping (Gielen and Houk, 1987; Wu et al., 1990).

In equilibrium point theories of motor control, when two opposing muscles are activated to a sufficient degree, the springlike dynamics of the muscle pair creates a potential well that contains a unique joint equilibrium position (Bizzi et al., 1982; Feldman, 1966; Hogan, 1984). In these models, the establishment of the potential well results in a movement of the joint, which ultimately comes to rest at the equilibrium point. By altering the relative activation of the opposing muscles, it is possible to alter the location of the equilibrium. Thus, a movement from one point to another can be achieved by shifting the joint equilibrium position from the initial point to the target location (Gribble et al., 1998; Mussa-Ivaldi, 1997).

Gomi and Kawato (1996) have argued that in order to produce realistic human reach trajectories with realistic muscle dynamic parameters, it is necessary that the equilibrium trajectory take on a form that is as complex to compute as the inverse dynamics themselves. However, this analysis makes the assumption that a muscle's force response (with the stretch reflex intact) is simply a linear function of both the current length and stretch rate of the muscle. In contrast, muscle stretch studies in the cat soleus muscle (Nichols and Houk, 1976) and the human wrist (Gielen and Houk, 1984) suggest that the stretch reflex plays a critical role during lengthening not only in the maintenance of muscle stiffness (implicitly assumed in the equilibrium point models) but also in the production of a damping force that is nonlinearly related to the velocity of stretch (Houk, 1981). These ideas have been further elaborated in the models of Hasan (1983), Gielen and Houk (1987), and Wu et al. (1990), and they have also proven to be useful in artificial control problems (Chang et al., 1999; Wu et al., 1997).

Ghez and Martin (1982) have shown that both the antagonist and second agonist muscle bursts during reaching movements in the cat are suppressed when the limb is prevented from moving. These results imply a significant role of the stretch reflex in the production of these latter bursts. The head movement study of Hannaford and Stark (1985) indicates that the triphasic muscle burst pattern occurs primarily during rapid head movements (those that are nearly time-optimal). Subsequent modeling work shows that reflex action can be used to derive the timing and magnitude of the braking (antagonist) and clamping (second agonist) muscle bursts (Ramos and Stark, 1987; Ramos et al., 1989). They observe that, similar to our model, an occurrence of the second agonist burst is contingent on an overshoot by the joint of the final resting position. Although both models include a Hill-type nonlinearity for muscle shortening, the Hannaford and Stark (1985) model utilizes a linear response of the stretch receptor to velocity. Furthermore, the Hannaford and Stark (1985) model assumes that the stretch reflex is not involved in shaping muscle activation during the initial muscle burst. In contrast, our formulation of the interaction between descending command and reflex activation is more straightforward in that both components are assumed to contribute to muscle activation at all times.

Lin and Rymer (1998) recently examined the contribution of damping by the stretch reflex in a preparation composed of a single muscle and a simulated inertial load. They observed that inclusion of the stretch reflex led to lightly damped oscillations of the simulated mass and not the stiction-like behavior that is observed in our model. However, it is important to note that the stiction property requires the coactivation of a pair of opposing muscles and only holds near the equilibrium position (see "FPD Interaction of Multiple Muscles"). Without the second muscle, energy is removed only during one-half cycle of the oscillation (Fagg et al., 2002). Thus, multiple cycles will be required before the system comes to rest, which is consistent with the Lin and Rymer (1998) experimental results.

Our theory has much in common with equilibrium point theories of motor control since an equilibrium point is specified, but the effective movement endpoint depends on the equilibrium point in a complex way that involves the dynamics of the arm (with spinal contribution) in a neighborhood around the equilibrium point (Barto et al., 1999; Wu et al., 1990). A consequence of FPD-induced stiction behavior is that the central motor system may not simply rely on a shifting of the equilibrium position from the initial to the target position. Ghez (1979) and Ghez and Martin (1982) suggest that neural commands that control limb movements in the cat appear to be composed of a high-magnitude (pulse) component followed by a smaller step component (Gielen and Houk, 1986). Although analogous to the pulse-step commands that control rapid eye movements (Keller and Robinson, 1971; Robinson, 1975), our model relies on the stretch reflex to brake the ongoing movement rather than relying on the intrinsic viscous behavior of the arm. Furthermore, in our model, the pulse has the function of moving the joint out of the stiction region, allowing the joint to achieve a nontrivial velocity before it arrives at the new stiction region specified by the step command corresponding to the target.

Our model is similar to that of Karniel and Inbar (1997) in that it relies on the natural dynamics of the arm and muscles to achieve realistic kinematic trajectories while specifying the control in terms of a feedforward pulse-step motor command waveform. Such a representation constitutes a simple description of the time-varying motor command as compared with, for example, a continuous representation of torque output as a function of time (e.g., Katayama and Kawato, 1993). However, in the model of Karniel and Inbar (1997), the parameters are separately specified for the agonist and antagonist muscle bursts. In contrast, Lestienne (1979) suggests that these bursts are not separately planned but instead are planned as a unit. Because our model relies on the stretch reflex for the online production of the braking antagonist muscle burst, we are in fact providing an explicit mechanism for pairing agonist and antagonist bursts. When the limb is externally loaded (e.g., with viscous, inertial, or elastic loads as used in the experiments of Gottlieb [1996]), it will follow a different path through phase space. Because the reflex formulation is sensitive to a complex combination of muscle length and stretch velocity, this will affect the magnitude and timing of the reflex-mediated antagonist response, even if the descending control parameters relating to the braking phase (co-contraction and step in our pulse-step formulation) are identical across the different loading conditions. Hence, the pairing process is not fixed and can be sensitive to external factors.

Karniel and Inbar (1999) have since explored the use of a simplified model of fractional power damping of the stretch reflex in braking of the ongoing movement. They show that this gives rise to the experimentally observed relationship among movement duration, movement amplitude, and peak velocity (Hanneton et al., 1997). Although our model (as well as that of Karniel and Inbar) utilizes a pulse-step formulation of the motor command, it is not our intention to posit this formulation as a complete theory of arm movement control. Instead, we wish to demonstrate on a qualitative level that the kinematics of reach can be roughly accounted for by assuming a combination of realistic motor plant dynamics, reflex circuitry, and a simple motor command.

Our results call into question theories that make extensive use of highly detailed motor plans that must be computed prior to reach initiation. In a more complete theory, we imagine allowing a constrained increase in complexity from the pulse-step waveform, as well as online adjustment of the motor signals as a function of delayed sensory feedback and motor efference copy. The work of Mussa-Ivaldi (1997) provides one hint as to how to approach the former. In his work, more complex motor signals are achieved by specifying individual pulse-step parameters over a basis set of Gaussian force fields, which roughly correspond to the activation of combinations of muscles. The latter issue we have addressed in the context of a simplified controlled system (Barto et al., 1999). We plan to return to these issues with the more realistic arm/muscle/spinal system presented here.

Acknowledgments

Preparation of this manuscript was supported by National Institutes of Health grant NIH MH 48185-09 and National Science Foundation grant EIA 9703217. The authors wish to thank Amir Karniel, Zev Rymer, Michael Kositsky, Mark Shapiro, and Richard Nichols for their valuable comments on this manuscript. The authors also thank Leo Zelevinsky and Anders Jonsson for their work on early versions of this model.

References

Amis, A.A., Dowson, D., and Wright, V. (1979) Muscle strengths and musculo-skeletal geometry of the upper limb. *Engineering in Medicine* 8(1): 41-48.

An, K.N., Hui, F.C., Morrey, B.F., Linscheid, R.L., and Chao, E.Y. (1981). Muscles across the elbow joint: A biomechanical analysis. *J Biomech* 14(10): 659-669.

Asatryan, D.G., and Feldman, A.G. (1965) Functional tuning of nervous system with control of movement or maintenance of a steady posture: I. Mechanographic analysis of the work of the joint on execution of a postural task. *Biophys USSR* 10: 925-935.

Barto, A.G., Fagg, A.H., Sitkoff, N., and Houk, J.C. (1999) A cerebellar model of timing and prediction in the control of reaching. *Neural Comput* 11: 565-594.

Binder, M.D., Heckman, C.J., and Powers, R.K. (1996) The physiological control of motorneuron activity. In Rowell, L.B., and Shepherd, J.T. (Eds.), *Handbook of physiology*, chapter 1. New York: Oxford University Press.

Bizzi, E., Chapple, W., and Hogan, N. (1982) Mechanical properties of muscles: Implications for motor control. *Trends Neurosci* 5: 395-398.

Bizzi, E., Hogan, N., Mussa-Ivaldi, F.A., and Giszter, S. (1992) Does the nervous system use equilibrium-point control to guide single and multiple joint movements? *Behav Brain Sci* 15: 603-613.

Chang, S.L., Wu, C.H., and Lee, D.T. (1999) A muscular-like compliance control for active vehicle suspension. In *Proc. 1999 IEEE International Conference on Robotics and Automation*, pp. 3275-3280. Detroit: IEEE.

Engelbrecht, S.E. (2001) Minimum principles in motor control. *J Math Psychol* 45: 497-542.

Fagg, A.H. (2000) A model of muscle geometry for a two degree-of-freedom planar arm. Technical Report 00-03, Department of Computer Science, University of Massachusetts, Amherst. **http://www-anw.cs.umass.edu/~fagg/papers/2000/muscle.html.**

Fagg, A.H., Houk, J.C., and Barto, A.G. (2002) Fractional power damping model of control. III Reflex responses to variations in mechanical load. Technical Report 02-7. Department of Computer Science, University of Massachusetts, Amherst.

Feldman, A.G. (1966) Functional tuning of the nervous system with control of movement or maintenance of a steady posture: II. Controllable parameters of the muscle. *Biophysics* 11: 565-578.

Feldman, A.G., Adamovich, S.V., Ostry, D.J., and Flanagan, J.R. (1990) The origin of electromyograms: Explanations based on the equilibrium point hypothesis. In Winters, J.M., and Woo, S.L.-Y. (Eds.), *Multiple muscle systems: Biomechanics and movement organization*, pp. 195-213. New York: Springer-Verlag.

Flanagan, J.R., Ostry, D.J., and Feldman, A.G. (1993) Control of trajectory modifications in target-directed reaching. *J Motor Behav* 25: 140-152.

Freivalds, A. (1985) Incorporation of active elements into the articulated total body model. Technical Report AAMRL-TR-85-061, Armstrong Aerospace Medical Research Laboratory, Wright-Patterson Air Force Base.

Georgopoulos, A.P., and Grillner, S. (1989) Visuomotor coordination in reaching and locomotion. *Science* 245: 1209-1210.

Ghez, C. (1979) Contributions of central programs to rapid limb movement in the cat. In Asanuma, H., and Wilson, V.J. (Eds.), *Integration in the nervous system,* pp. 305-320. Tokyo: Igaku-Shoin.

Ghez, C., and Martin, J.H. (1982) The control of rapid limb movement in the cat: III. Agonist-antagonist coupling. *Exp Brain Res* 45: 115-125.

Gielen, C.C.A.M., and Houk, J.C. (1984) Nonlinear viscosity of human wrist. *J Neurophysiol* 52: 553-569.

Gielen, C.C.A.M., and Houk, J.C. (1986) Simple changes in reflex threshold cannot explain all aspects of rapid voluntary movements. *Behav Brain Sci* 9: 605-607.

Gielen, C.C.A.M., and Houk, J.C. (1987). A model of the motor servo: Incorporating nonlinear spindle receptor and muscle mechanical properties. *Biol Cybern* 57: 217-231.

Gomi, H., and Kawato, M. (1996) Equilibrium-point control hypothesis examined by measured arm stiffness during multijoint movement. *Science* 272 (5): 117-120.

Gottlieb, G.L. (1996) On the voluntary movement of compliant (inertial-viscoelastic) loads by parcellated control mechanisms. *J Neurophysiol* 76(5): 3207-3229.

Gribble, P.L., Ostry, D.J., Sanguineti, V., and Laboissiere, R. (1998) Are arm control signals required for human arm movement? *J Neurophysiol* 79(3): 1409-1424.

Hannaford, B., and Stark, L. (1985) Roles of the elements of the triphasic control signal. *Exp Neurol* 90: 619-634.

Hanneton, S., Berthoz, A., Droulez, J., and Slotine, J.J.E. (1997) Does the brain use sliding variables for the control of movements? *Biol Cybern* 73: 381-393.

Hasan, Z. (1983) A model of spindle afferent response to muscle stretch. *J Neurophysiol* 49(4): 989-1006.

Hasan, Z., and Houk, J.C. (1975) Transition in sensitivity of spindle receptors that occurs when muscle is stretched more than a fraction of a millimeter. *J Neurophysiol* 38: 673-689.

Hill, A.V. (1938) The heat of shortening and the dynamic constants of muscle. *Proc Roy Soc (Lond), ser B* 126: 136-195.

Hogan, N. (1984) Adaptive control of mechanical impedance by coactivation of antagonist muscles. *IEEE Trans Auto Cont* AC-29: 681-690.

Hogan, N., Bizzi, E., Mussa-Ivaldi, F., and Flash, T. (1987) Controlling multijoint motor behavior. *Exerc Sport Sci Rev* 15: 153-190.

Hollerbach, J., and Flash, T. (1982) Dynamic interactions between limb segments during planar arm movement. *Biol Cybern* 44: 67-77.

Houk, J.C. (1981) Afferent mechanisms mediating autogenetic reflexes. In Pompeiano, O., and Marsan, C.A. (Eds.), *Brain mechanisms of perceptual awareness,* pp. 167-181. New York: Raven.

Houk, J.C., Crago, P.E., and Rymer, W.Z. (1981) Function of the spindle dynamic response in stiffness regulation—a predictive mechanism provided by non-linear feedback. In Taylor, H., and Prochazka, A. (Eds.), *Muscle receptors and movement,* pp. 299-309. London: Macmillan.

Houk, J.C., and Rymer, W.Z. (1981) Neural control of muscle length and tension. In Brooks, V.B. (Ed.), *Handbook of physiology,* sec. 1, vol. 2, Motor control, pp. 247-323. Bethesda, MD: American Physiological Society.

Houk, J.C., Rymer, W.Z., and Crago, R.E. (1992) Responses of muscle spindle receptors to transitions in stretch velocity. In Jami, L., Pierrot-Deseilligny, E., and Zytnicki, D. (Eds.), IBRO Symposium Series, *Proc. of the Symposium: Muscle Afferents and Spinal Control of Movement,* September 16-19, 1991, College de France, Paris, pp. 53-61. New York: Pergamon Press.

Houk, J.C., Singer, J.J., and Goldman, M.R. (1970) An evaluation of length and force feedback to soleus muscles of decerebrate cats. *J Neurophysiol* 33: 784-811.

Jordan, M.I., Flash, T., and Arnon, Y. (1994) A model of the learning of arm trajectories from spatial deviations. *J Cognitive Neurosci* 6: 359-376.

Karniel, A., and Inbar, G.F. (1997) A model of learning human reaching movements. *Biol Cybern* 77: 173-183.

Karniel, A., and Inbar, G.F. (1999) The use of a nonlinear muscle model in explaining the relationship between duration, amplitude, and peak velocity. *J Mot Behav* 31(3): 203-206.

Katayama, M., and Kawato, M. (1993) Virtual trajectory and stiffness ellipse during multi-joint arm movement predicted by neural inverse models. *Biol Cybern* 69: 353-362.

Keller, E.L., and Robinson, D.A. (1971) Absence of stretch reflex in extraocular muscles of the monkey. *J Neurophysiol* 34: 908-919.

Lestienne, F. (1979) Effects of inertial load and velocity on the braking process of voluntary limb movements. *Exp Brain Res* 35: 407-418.

Lin, D.C., and Rymer, W.Z. (1998) Damping in reflexively active and areflexive lengthening muscle evaluated with inertial loads. *J Neurophysiol* 80(6): 3369-3372.

Lukashin, A.V., Amirikian, B.R., and Georgopoulos, A.P. (1996) Neural computations underlying the exertion of force: A model. *Biol Cybern* 74: 469-478.

Massone, L.L.E., and Myers, J.D. (1996) The role of plant properties in arm trajectory formation: A neural network study. *IEEE Trans Syst Man Cybern* 26: 719-732.

Miller, L.E., and Sinkjaer, T. (1998) Primate red nucleus discharge encodes the dynamics of limb muscle activity. *J Neurophysiol* 80: 59-70.

Mussa-Ivaldi, F.A. (1992) From basis functions to basis fields: Vector field approximation from sparse data. *Biol Cybern* 67: 479-489.

Mussa-Ivaldi, F.A. (1997) Nonlinear force fields: A distributed system of control primitives for representing and learning movements. In *Proc. of the IEEE International Symposium on Computational Intelligence in Robotics and Automation–CIRA 1997*, pp. 84-90. Los Alamitos, CA: IEEE Computer Society Press.

Nichols, T.R., and Houk, J.C. (1976) Improvement in linearity and regulation of stiffness that results from actions of stretch reflex. *J Neurophysiol* 39(1): 119-142.

Partridge, L.D. (1979) Muscle properties: A problem for the motor controller physiologist. In Talbott, R.E., and Humphries, D.R. (Eds.), *Posture and movement*, pp. 189-229. New York: Raven Press.

Peterson, B., Baker, J.F., Iwamoto, Y., Perlmutter, S.I., and Quinn, K. (1992) Neuronal substrates of vestibular reflex control of eye and head movements. In Shimazu, H., and Shinoda, Y. (Eds.), *Vestibular and brain stem control of eye, head and body movements*, pp. 53-68. Tokyo: Japan Scientific Societies Press.

Ramos, C.F., and Stark, L.W. (1987) Simulation studies of descending and reflex control of fast movements. *J Mot Behav* 19(1): 38-61.

Ramos, C., Stark, L., and Hannaford, B. (1989) Time optimality, proprioception, and the triphasic EMG pattern. *Behav Brain Sci* 12: 231-232.

Robinson, D.A. (1975) Oculomotor control signals. In Lennerstrand, G., and Bach-y-rita, P. (Eds.), *Basic mechanisms of ocular motility and their clinical implications*, pp. 337-374. Oxford: Pergamon Press.

van Dijk, J.H.M. (1978) Simulation of human arm movements controlled by peripheral feedback. *Biol Cybern* 29: 175-186.

van Sonderen, J.F., and Denier van der Gon, J.J. (1990) A simulation study of a programme generator for centrally programmed fast two-joint arm movements: Responses to single- and double-step target displacements. *Biol Cybern* 63: 35-44.

Winters, J.M. (1990) Hill-based muscle models: A systems engineering perspective. In Winters, J.M., and Woo, S.L.-Y. (Eds.), *Multiple muscle systems: Biomechanics and movement organization*, pp. 69-93. New York: Springer-Verlag.

Winters, J.M., and Stark, L. (1985) Analysis of fundamental human movement patterns through the use of in-depth antagonist muscle models. *IEEE Trans Biomed Eng* 32: 826-839.

Winters, J.M., and Stark, L. (1987) Muscle models: What is gained and what is lost by varying model complexity. *Biol Cybern* 55: 403-420.

Winters, J.M., and Stark, L. (1988) Estimated mechanical properties of synergistic muscles involved in movements of a variety of human joints. *J Biomechanics* 21: 1027-1041.

Wu, C.H., Houk, J.C., Young, K.Y., and Miller, L.E. (1990) Nonlinear damping of limb motion. In Winters, J.M., and Woo, S.L.-Y. (Eds.), *Multiple muscle systems: Biomechanics and movement organization*, pp. 214-235. New York: Springer-Verlag.

Wu, C.H., Hwang, K.S., and Chang, S.L. (1997) Analysis and implementation of a neuromuscular-like control for robotic compliance. *IEEE Trans Cont Syst Tech* 5(6): 586-597.

Yamaguchi, G.T., Sawa, A.G., Moran, D.W., Fessler, M.J., and Winters, J.M. (1990) A survey of human musculotendon actuator parameters. In Winters, J.M., and Woo, S.L.-Y. (Eds.), *Multiple muscle systems: Biomechanics and movement organization*, pp. 717-773. New York: Springer-Verlag.

Appendix

Muscle Stiffness

We assume that muscle force range in equation (8.2) is proportional to physiological cross-sectional area. We use the PCSA values summarized in Yamaguchi et al. (1990); the original references are given in table 8.1. Combining this with equation (8.2) and assuming that the normalized stiffness (K_n) is equal to 1:

$$K_m = \frac{Q\,PSCA_m}{l_m^r}, \qquad (8.10)$$

where Q is a constant that describes the transformation from physiological cross-sectional area to maximum force and l_m^r is the length range of muscle m over all feasible skeletal configurations. The l_m^r for each muscle is derived from our model of muscle geometry and given in table 8.1.

Astryan and Feldman (1965) estimate the joint stiffness of the elbow in a task in which a steady posture was first established against a force at a given joint position (corresponding to $\alpha_6^{(1)}$, $\alpha_6^{(2)}$ and $\alpha_6^{(3)}$ in figure 8.6b). The force was then suddenly reduced, resulting in a further flexion of the elbow. If we assume that the subjects did not explicitly react to the change in force (as they were instructed), then the joint position at which the elbow came to rest after the force reduction ($\alpha_5^{(*)}$ to $\alpha_{-45}^{(*)}$) can be interpreted as the point at which the external force is exactly balanced by the tonic stretch reflex for the motor command that was originally established *before* reduction of the external force. Thus, the vertically running curves in figure 8.6b ($M_{(\psi)}^{(*)}$) can be interpreted as the tonic position-torque response of the muscles as a function of three different motor command magnitudes.

In computing an estimate of tonic stretch reflex gain, we assume that $\alpha_6^{(2)}$ to $\alpha_3^{(2)}$ represents the linear stiffness region for the elbow flexors. A change in load resisted by the elbow of -0.42 kg/m , or -4.12 N/m, results in a change in elbow position by $8.5°$. We assume that the biceps and brachialis are the two muscle groups primarily involved in resisting the external torque. In our model of muscle geometry, these two muscles reduce their muscle length over this range by 4.5 mm (0.18 in.) and 4.9 mm (0.19 in.), respectively. We also assume that the moment arms are not changing within this region; we therefore arrive at the following relation:

$$\Delta\tau_e = \Delta F_{brac}R_{e,brac} + \Delta F_{bicep}R_{e,bicep}.$$

Furthermore, if we assume that the moment arms for both muscles are approximately the same (this is a reasonable assumption given our model of muscle geometry: $R_{e,brac} = 3.24$ cm [1.28 in.] and $R_{e,bicep} = 3.47$ cm [1.37 in.]), then:

$$\Delta\tau_e \approx \left(\frac{R_{e,brac} + R_{e,bicep}}{2}\right)\left(\Delta F_{brac} + \Delta F_{bicep}\right).$$

Incorporating the linear assumption of stiffness and equation (8.10):

$$\Delta\tau_e \approx \left(\frac{R_{e,brac} + R_{e,bicep}}{2}\right)\left(\Delta l_{brac}K_{brac} + \Delta l_{bicep}K_{bicep}\right);$$

$$= \left(\frac{R_{e,brac} + R_{e,bicep}}{2}\right)\left(Q\Delta l_{brac}\frac{PCSA_{brac}}{l_{brac}^r} + Q\Delta l_{bicep}\frac{PCSA_{bicep}}{l_{bicep}^r}\right).$$

Finally, solving for Q, we arrive at the following:

$$Q = \frac{2\Delta\tau_e}{R_{e,brac} + R_{e,bicep}}\left(\frac{1}{\dfrac{\Delta l_{brac}PCSA_{brac}}{l_{brac}^r} + \dfrac{\Delta l_{bicep}PCSA_{bicep}}{l_{bicep}^r}}\right).$$

In comparing the tonic response of the stretch reflex following either the lengthening or shortening of an active muscle, Gielen and Houk (1984) observed a lower stiffness following a lengthening event. Since we are primarily interested here in the lengthening case, we choose the following estimate of Q to compensate for the shortening condition used by Astryan and Feldman (1965):

$$Q = 0.75\frac{2\Delta\tau_e}{R_{e,brac} + R_{e,bicep}}\left(\frac{1}{\dfrac{\Delta l_{brac}PCSA_{brac}}{l_{brac}^r} + \dfrac{\Delta l_{bicep}PCSA_{bicep}}{l_{bicep}^r}}\right);$$

$$= 202.87.$$

The resulting muscle reflex values are summarized in table 8.1.

Extent of Muscle Exponential Region

The extent of the exponential region of each muscle is determined by parameter c in equation (8.1). We assume that each muscle's c_m is linearly related to the length range of the muscle. Specifically:

$$c_m = C\, l_m^r.$$

We assume that the exponential region of the muscle length-force relationship in Feldman (1966) is from $\alpha_0^{(2)}$ (88°) to $\alpha_3^{(2)}$ (111.5°) (see figure 8.6b). By this we mean that the exponential term of equation (8.1) saturates over this range. Thus:

$$c_m \approx \left(1 - e^{-1}\right)\Delta\hat{l}_m,$$

where $\Delta\hat{l}_m$ is the change in muscle length over the exponential region. For the elbow flexors, $\hat{l}_{brac} = 2.52$ cm (0.99 in.) and $\hat{l}_{bicep} = 2.62$ cm (1.03 in.).

Utilizing the average length change and length range of the brachialis and biceps, we estimate C as follows:

$$C_m \approx \left(1 - e^{-1}\right)\frac{\Delta\hat{l}_{brac} + \Delta\hat{l}_{bicep}}{l_{brac}^r + l_{bicep}^r};$$

$$= 10.23.$$

Thus, $c_{brac} = 0.91$ cm (0.36 in.) and $c_{bicep} = 2.34$ cm (0.92 in.). The remaining parameters are given in table 8.1.

Stiffness of Flexor Carpi Radialis

An et al. (1981) report the PCSA of flexor carpi radialis as 2.0 cm² (0.31 in.²). Given equation (8.10), we arrive at $K_{fcr} = 8,610$ N/m.

9

The Multidimensional and Temporal Regulation of Limb Mechanics by Spinal Circuits

T.R. Nichols and R.J.H. Wilmink
Department of Physiology,
Emory University

T.J. Burkholder
Department of Applied Physiology,
Georgia Institute of Technology

For a broad range of motor tasks, including standing and locomotion, postural regulation is required to maintain balance and stability in the face of external forces and internally generated disturbances. This regulatory system is distributed in both space and time to meet the needs of a linked musculoskeletal system. Recent evidence strongly implicates neural feedback from muscle spindles and Golgi tendon organs in this ongoing regulation. In this chapter, we argue that short-latency pathways from these muscle proprioceptors contribute to this regulation through convergent force and length feedback.

Macpherson (1988a, 1988b, 1994) has investigated the coordinated responses of standing cats to postural disturbances. During standing with normal limb spacing, the limbs and trunk responded to horizontally directed perturbations in such a way as to oppose the disturbance to the animal's center of mass (Macpherson, 1988a). Within each limb, however, the ground reaction force vectors were aligned predominantly along two opposite directions. Modulation according to direction occurred by variations in the magnitudes of these preferred vectors rather than by direction. After removal of the vestibular apparatus, this strategy was largely intact, suggesting an important role for proprioceptive feedback (Thomson et al., 1991). This strategy could have resulted from the convergence of proprioceptive feedback onto a brainstem area that would compute the appropriate activation patterns of muscles in the four limbs. Alternatively, the appropriate responses and resulting global response vector could have resulted from computations in a distributed propriospinal network in the spinal cord.

During locomotion, selective removal of proprioceptive feedback from the triceps surae muscles leads to deficits in postural regulation (Abelew et al., 2000; Nichols, Cope, and Abelew, 1999) during both standing and locomotion. Proprioceptive feedback was removed selectively from the triceps surae muscles of the right hindlimb of cats by transecting the nerves to those muscles and allowing self-reinnervation (Cope, Bonasera, and Nichols, 1994; Cope and Clark, 1993). When the animals walked on a level surface or up a ramp, interjoint coordination was within normal limits for both hindlimbs; that is, the joints of the hindlimb tended to flex and extend together. When the animals walked down a ramp, however, a pronounced yield occurred at the ankle joint during the stance phase of the step cycle in the limbs containing the reinnervated muscles, and a breakdown in coordination between ankle and knee resulted.

The apparent increase in compliance of the ankle in the treated limb was correlated with the functional role of the musculature. During level and uphill walking in normal and treated animals, the triceps surae muscles shortened during contraction for most of the stance phase. During downhill walking, these muscles normally underwent substantial periods of lengthening contraction, as they provide a braking action for this task. It is during this phase of the step cycle that the breakdown in coordination occurred in the treated animals. These data indicate that short-latency, proprioceptive feedback from muscles is important in managing motor coordination during conditions of active lengthening.

These examples indicate that proprioceptive feedback contributes to motor coordination during both standing and locomotion. A detailed interpretation of these experiments and knowledge of proprioceptive mechanisms can provide important clues concerning the identities of the involved receptors and the organizational features of the feedback pathways. The deficits of coordination during locomotion produced by the loss of local feedback by reinnervation are most readily explained by disruption of pathways from muscle spindle receptors back to the same or synergistic muscles. The organization of these linkages suggests that pathways from muscle spindle receptors subserve coordination of muscles that cross a given joint in response to disturbances (Lloyd, 1946; Nichols, 1994). Although excitatory connections from muscle spindle receptors across joints are known to exist (Eccles et al., 1957a; Edgley et al., 1986), these pathways appear to be substantially weaker than autogenic pathways or those linking muscles of a synergistic group (Wilmink, 1998). Therefore, the autogenic pathways and pathways to synergistic muscles are probably more important in mediating coordination during stance.

In contrast, feedback from Golgi tendon organs is distributed more widely across joints and among groups of synergistic muscles (Eccles et al., 1957b; Nichols, 1994; Nichols et al., 1999; Wilmink, 1998). Autogenic connections from group Ib fibers do exist (Watt et al., 1976), but these pathways appear to be relatively weak (Rymer and Hasan, 1980; but see Kirsch and Rymer, 1992). A given motor nucleus generally receives input from group Ib afferents from several different muscles. Therefore, disruption of force feedback from only the triceps surae muscles might not lead to a detectable deficit because of the diffuse distribution of these pathways. During locomotion, excitatory, polysynaptic pathways from muscle spindle and tendon organ afferents ("group I excitation") are expressed that appear to be widely distributed among the antigravity muscles of the limb (Guertin et al., 1995; Prochazka, 1996). Since these pathways appear to be even more widely distributed than those of negative force feedback (Nichols, 1994), withdrawal of such feedback from one muscle group would not be expected to result in a localized disruption of coordination. These considerations support the hypothesis that the loss of coordination during downhill walking is due mainly to the removal of local feedback from muscle spindle receptors (Nichols et al., 1999).

In addition to differences in spatial distribution of feedback from muscle spindles and Golgi tendon organs, the dynamic properties of both types of feedback differ also. These dynamic characteristics can originate in the properties of both individual receptors and associated synaptic pathways. Reflex action from primary muscle spindle receptors is dependent upon the magnitude of length change, movement dynamics, and length history (Houk et al., 1973, 1981b; Huyghues-Despointes, 1998; Lin and Rymer, 1993; Nichols et al., 1999; Proske et al., 1992). These characteristics are similar to the nonlinear mechanical properties of muscle. When these elements are combined in the stretch reflex, simplified behavior results (Nichols and Houk, 1976).

In contrast to the complex manner in which muscle spindle receptors respond to length changes, force feedback from Golgi tendon organs appears to be less dependent upon amplitude and dynamic characteristics of the mechanical disturbances. However, complex properties of force feedback are introduced at the level of interneuronal pathways. For example, force feedback appears to be considerably more prolonged than length feedback, at least in the decerebrate state (Nichols, 1999). Furthermore, a phenomenon of "inhibitory fading" has been reported (Zytnicki et al., 1990) during electrical stimulation of a muscle, despite continuing response from the receptors themselves. Inhibitory fading is not observed during natural stimulation (Hayward et al., 1988; Nichols, 1999), so the physiological significance of this phenomenon has yet to be determined. Finally, the active set of pathways mediating force feedback can change as a function of behavioral state (Pearson, 1995; Prochazka, 1996). The expression of "group I excitation" during locomotion and fictive locomotion (Guertin et al., 1995) clearly represents a major reorganization of proprioceptive feedback.

In this chapter, we consider the manner in which force feedback and length feedback are combined in the spinal cord to achieve motor coordination. Quantitative models combining the actions of length and force feedback have been few. The first such model subjected to quantitative evaluation was the original stiffness regulation hypothesis (Houk, 1972, 1979; Nichols and Houk, 1976). According to this hypothesis of autogenic reflex action, length and force feedback would summate postsynaptically at the motoneuron membrane. The advantage of this scheme was that the mechanical characteristics of the muscle could be modulated without changing the net gain of feedback. Muscular stiffness, rather than being determined by the intrinsic stiffnesses of progressively recruited motor units, would be regulated by combined length and force feedback. Stiffness would be increased by a proportionately larger gain of length feedback, whereas compliance would be enhanced by a proportionately larger gain of force feedback. The two pathways could therefore be modulated in a reciprocal fashion to alter muscular stiffness. Internal disturbances such as fatigue, however, would be equally well compensated in all cases since the total feedback gain could be preserved. If stiffness regulation were mediated by length feedback only, muscular stiffness could be modulated simply by altering the gain of length feedback. However, reductions in gain would result in poorer compensation for the nonlinear properties of muscle.

Although strong evidence showed that muscular stiffness is indeed regulated by autogenic reflexes (Nichols and Houk, 1976), evidence for a role of force feedback showed this pathway to be rather weak in decerebrate cats (Houk et al., 1970; Rymer and Hasan, 1980). Stiffness regulation for a given muscle was therefore attributed mainly to the actions of the primary receptor of the muscle spindle (Houk et al., 1981a). The major characteristics of this control system were observed by evaluating the mechanical properties of reflexive and areflexive muscle at different forces and amplitudes of stretch. Reflex responses are large at low forces and decline toward higher forces. This dependence complements the increase in intrinsic muscular stiffness with force so that total stiffness is less sensitive to background force (Hoffer and Andreassen, 1981; Nichols and Houk, 1976). Muscular stiffness is also less dependent on the amplitude of length change in the reflexive than in the areflexive state (Nichols and Houk, 1976). Since the stiffness of reflexive muscle is less dependent upon force and amplitude than is intrinsic muscular stiffness, length feedback renders the muscle more springlike.

Although a role for force feedback in autogenic reflex action is unresolved, the integration of force with length feedback clearly occurs in the system of intermuscular, or heterogenic, reflexes (Nichols, 1994, 1999; Wilmink, 1998). This feedback is dependent upon the spatial distribution and mechanical coupling of the associated muscles, the background forces and lengths of the muscles, and the temporal characteristics of mechanical perturbations. Despite the complexities of this network of pathways, force feedback should in principle increase total limb compliance and increase interjoint coordination. Because of these actions, we propose that this network regulates stiffness (or compliance) of the limb. Since interjoint coordination is necessary to maintain muscles at lengths favoring full force-generating capacities, the control of limb mechanics should be most effective when interjoint coordination is maintained. In this chapter, we provide an update of the spatial and temporal aspects of force and length feedback and then an evaluation of the hypothesis that integrated force and length feedback governs resultant limb stiffness and interjoint coordination.

Musculoskeletal Mechanics and the Spatial Distribution of Length Feedback

An analysis of the structure and mechanics of the musculoskeletal system provides a basis for understanding the actions of distributed feedback (figure 9.1). The intrinsic mechanics of the passive hindlimb contribute to its behavior through two mechanisms: the nonuniform distribution of inertia and the mechanical constraints of connective tissue. The inertial properties of the limb segments act as a mechanical filter, and the inertia of the hindlimb is not uniformly distributed: the major moment of inertia of the foot is barely 6% that of the thigh (Hoy and Zernicke, 1985). This nonuniform distribution of inertia, while an effective mechanism for increasing locomotor efficiency, makes the distal segments much more sensitive to mechanical perturbations. The passive tissues of the skeletal system (i.e., bone, cartilage, and ligaments) act as springs to constrain motion of the limb. These tissues constrain the joints in a way that is often assumed to approximate simple hinges. Although the motion at the knee and ankle can largely be described as a parasagittal plane rotation about a single axis, much greater mechanical fidelity can be achieved by considering both joints as compound rotations within near-sagittal and near-transverse planes (Burkholder and Nichols, 2000; Hollister et al., 1992). At the knee, the collateral ligaments strongly couple adduction with internal rotation, justifying the reduction to two degrees of freedom (DOF). Under the compressive loading of muscle action and weight support, the bony structure of the talocrural articulation provides a similar coupling at the ankle.

Some generalizations can be made from these observations. The inertial properties of the passive limb will tend to reduce the effect of any perturbation on the more proximal limb segments. This can contribute to increased body stability but may increase the risk of disproportionate extension or flexion of the distal joints, particularly the ankle. It is advantageous to distribute a perturbation more uniformly across the joints of the limb so that muscles can work within ranges of length that maximize force generation and moment arm (Engberg and Lundberg, 1969; Goslow et al., 1973). Another consequence of segmented limbs is that the response to an applied force perturbation may not act in a direction that is opposite to the direction of the perturbation (Hogan, 1985; Mussa-Ivaldi et al., 1985). The relationship between the

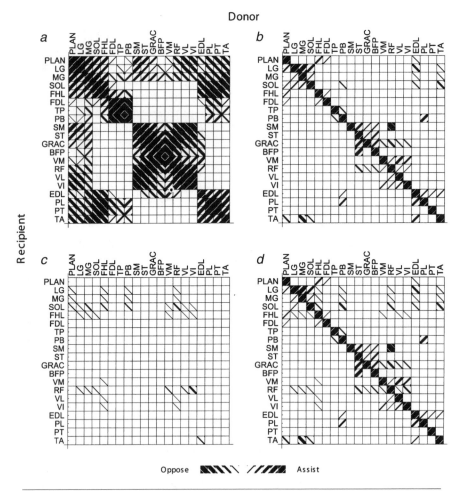

Donor

Oppose ■■■▨▨▨ ▨▨▨▨■ Assist

Figure 9.1 Mechanical and neural interactions among hindlimb muscles. *(a)* Mechanical similarity of muscle actions. Cells are coded according to the cosine of the angle between the lines of action of each pair of muscles, with increasing thickness of frontslash indicating increasingly similar actions (assist) and increasing thickness of backslash indicating increasingly opposing (oppose) actions. For example, to compare, the medial and lateral gastrocnemius (MG and LG) both flex the knee and extend and abduct the ankle, but they have opposing abduction or adduction actions at the knee. The angle between these compound actions is 15°. *(b)* Magnitude of length feedback between muscle pairs. Increasing thickness of frontslash indicates increasing excitation, whereas increasing thickness of backslash indicates increasing inhibition (e.g., feedback from MG onto LG is strongly excitatory, whereas feedback from LG onto MG is comparatively weak). (Based on data from Bonasera and Nichols, 1994, 1996; Eccles et al., 1957a, 1957b; Nichols, 1999.) *(c)* Force feedback. Increasing thickness of backslash indicates increasing force-dependent inhibition. (Based on data from Bonasera and Nichols, 1994; Wilmink, 1998) *(d)* Sum of length and force feedback. Sum of *(b)* and *(c)*, representative of total proprioceptive feedback between pairs of muscles.

direction of applied perturbation and the direction of the force response will be further influenced by the orientation of the axes of rotation of the biaxial joints and the arrangement of connective tissues. We argue later that reflex mechanisms as well as the coupling provided by multiarticular and multiaxial muscles may mediate this interjoint coordination.

The intrinsic mechanical behavior of the muscular system has some capacity to distribute a perturbation across joints and orient the force response. Perturbation of the limb results in changes in the length of each muscle. Without any neural modulation, those active muscles that are stretched by the perturbation respond with increased force, whereas those that are shortened produce reduced force. Because of the distribution of segment inertias and mechanical filtering of the limb, a perturbation evokes the largest movement at the most distal segment. Biarticular muscles, in responding to this perturbation, increase the perturbing torque at the more proximal joints. For example, when the toe is vertically displaced, the ankle flexes, stretching the triceps surae muscles. The biarticular gastrocnemius muscles generate a flexion torque at the knee, reducing the amount of motion required at the ankle. Similarly, as the knee flexes, stretching the rectus femoris, a hip flexion torque is generated, reducing the motion required at the knee.

In the interest of regulating the posture of the body as a whole, however, it may not be generally appropriate for the mechanical response of the limb to result solely from the intrinsic mechanics of the musculature and limb architecture. If each joint is modeled as two coupled hinges, each muscle can have actions about both hinges of each joint it crosses. This implies that perturbations with respect to one axis will result in torque changes at the other axis as well. For example, the medial gastrocnemius (MG) muscle both extends and abducts the ankle. When the ankle is forcibly flexed, the MG is stretched and generates increased force and a restoring extension torque. This muscle also generates an increased abduction torque, which will tend to displace the joint in that direction. The net effect of this complex reaction may not be the appropriate response for regulating the posture of the body as a whole. In addition, the mechanical contributions of each muscle will vary according to limb orientation (Lawrence and Nichols, 1999a, 1999b). Furthermore, the three-dimensional responses of the intact limb to perturbations will depend upon the activation levels of the associated musculature since the intrinsic force responses of each muscle depend directly on activation level (Houk and Rymer, 1981). Changes in the distribution of activity among a set of muscles would lead to an alteration of the direction of the net response vector since the direction of the response vector depends upon the magnitude of each of the component vectors. Finally, more proximal muscles that cross more than one joint and have long fibers will maintain full force-generating capacity over a broader range of limb orientations and joint angles than more distal muscles with shorter fibers. The joint torques exerted by the more distal muscles will be more affected by joint angle due to their length-tension characteristics and variation in moment arms. These complexities directly influence the intrinsic mechanical properties of the musculoskeletal system and therefore impose important constraints on postural responses.

Neuronal mechanisms select the combination of muscles with the appropriate spatial characteristics to mediate postural regulation of the body. The first neuronal mechanism by which the restoring forces of the limb can be altered consists of the nonuniform distribution of the strengths of the autogenic reflexes of the component muscles. The limb is organized so that the majority of muscle mass and torque-generating capacity are aligned with the principal joint motion. The ground reaction forces of the triceps surae muscles appear to be directed in such a way as to promote postural stability of the feline body (Lawrence et al., 1993; Nichols, 1994), but the transverse stability of the ankle depends upon relatively small stabilizing muscles (Bonasera and Nichols, 1996). The powerful stretch reflexes and mutual reciprocal inhibition shared by these stabilizing muscles compensate for the relative lack of muscle tissue (Burkholder and Nichols, 2000) and increase the contributions of these muscles to the mechanical responses of the ankle.

The second mechanism by which the nervous system improves the response to perturbations is by coupling the actions of multiple muscles. Length feedback from muscle spindles excites not only the homonymous muscle but also a range of muscles with similar mechanical actions (McCrea, 1986; Nichols et al., 1999). Further, spindle afferents generally inhibit muscles with opposing mechanical functions (Hultborn, 1976; Lloyd, 1946). This pattern of excitation of agonists and inhibition of antagonists is the foundation of the essentially one-dimensional myotatic unit (Lloyd, 1946). This myotatic unit concept, representing one of the first attempts to explain the distribution of heterogenic length feedback, viewed length feedback and associated inhibitory interneurons as a mechanism for integrating muscles acting at a joint. According to this concept, excitation from primary receptors of muscle spindles reinforces the activation of synergistic muscles, and reciprocal inhibition supports the suppression of antagonistic muscles during voluntary movement. This scheme, which matches spinal circuit organization to musculoskeletal architecture, was later used to explain how the mechanical properties of a joint might be regulated for maintaining posture (Houk and Henneman, 1974). This arrangement makes the action of heterogenic length feedback on a joint equivalent to autogenic length feedback on a muscle.

As mentioned earlier, joint motions take place in three dimensions, and few muscles exert identical actions with respect to either line of action or joints spanned (Eccles and Lundberg, 1958). Although length feedback generally links muscles that act in common or in opposition at a joint, the details of this distribution are more complex and reflect the mechanical relationships among the associated muscles. The mechanical interactions of the muscles of the limb at the knee and ankle are shown in figure 9.1a. The actions of each muscle at these two joints can be summarized by one moment arm vector (Burkholder and Nichols, 2000). The cosines of the angles between these vectors (figure 9.1a) form a symmetrical matrix. The organization of length feedback (figure 9.1b) is roughly similar, indicating that length feedback generally reinforces limb architecture. However, the distribution of heterogenic feedback does not incorporate all muscles with common actions and is frequently asymmetric with respect to a muscle pair. For example, there is little length feedback from the gastrocnemius muscles acting on the plantaris muscle, despite the nearly identical actions of the lateral gastrocnemius (LG) and plantaris at the ankle. Length feedback from MG acting on LG is much more powerful than that from LG acting on MG.

This nonuniform exchange of length feedback, combined with autogenic reflex action, produces a weighted average of the response vectors of a set of muscles that is not very sensitive to level of activation. This arrangement contrasts with an intrinsic response whose direction and magnitude are subject to the activation levels and lines of action of the muscles. When a joint is manipulated in different directions, the responses can be characterized by a "stiffness ellipse" (Mussa-Ivaldi et al., 1985). In the absence of length feedback, the "intrinsic" stiffness ellipse would change size and shape for different patterns of activation in the involved muscles. The regulatory properties of autogenic and heterogenic length feedback tend to reduce this dependence and therefore simplify the responses of the limb to perturbations.

A similar argument can be applied to coordination across joints. The relative stiffnesses of different joints tend to remain constant for different activation levels in the associated muscles in the presence of length feedback. In addition, the mechanical linkages between joints due to biarticular muscles remain strong, even at low force levels in the presence of reflex action. Therefore, the presence of length feedback promotes cross-joint and cross-axis coordination in the face of changing patterns of muscular activation.

In simulations of force generation by the distal segments of the hindlimb (i.e., the ankle and knee mobilized but the hip and above immobilized), the progressive addition of layers of length feedback has a dramatic effect on the direction and magnitude of endpoint force (Burkholder and Nichols, 2000). Under the action of the intrinsic properties of muscle alone or with the addition of autogenic or heterogenic length feedback, the knee-ankle assembly demonstrates a preferred path. Given an initial pattern of muscle activation, there is generally an anterior-posterior line along which the endpoint will tend to move. Transverse displacements or perturbations from this line generally result in opposing restoring forces. Under the action of intrinsic properties alone, this preferred path lies in a nearly parasagittal plane. Autogenic length feedback results in forces that point more medially, a pattern attributable to enhanced stiffness of the so-called stabilizing muscles of the ankle. Inclusion of heterogenic length feedback further augments this pattern, generally reinforcing the action of the autogenic length feedback.

The Spatial Distribution and Actions of Force Feedback

In contrast to the generally localized distribution of monosynaptic projections from primary receptors of muscle spindles, the oligosynaptic projections of Golgi tendon organs are more widely distributed. Early electrophysiological studies suggest that inhibitory pathways from Golgi tendon organs are distributed *universally* among extensor muscles of the limb (Eccles et al., 1957b; Laporte and Lloyd, 1952) and are mediated predominantly by disynaptic linkages (Eccles et al., 1957b; McCrea, 1986). Trisynaptic excitatory linkages were also found to extend from extensor muscles to flexor muscles. Eccles and his coworkers noted that, although this pattern of connectivity is reminiscent of the organization of flexor reflexes, it is unique in several respects, including the lack of inhibitory effects from flexor muscles to extensor muscles. Recent evidence supports the hypothesis that these pathways mediate the system of negative force feedback (Nichols, 1999).

Recent work has provided a more detailed picture of the distribution of force feedback among antigravity (extensor) muscles of the feline hindlimb. Force feedback appears not to be universally distributed across antigravity muscles. For example, strong inhibitory force feedback has been observed to extend from the gastrocnemius to soleus muscles but not between LG and MG (Nichols, 1989, 1999). These findings suggested that force feedback is distributed according to either motor unit fiber type or muscle articulation (Nichols, 1994). The results of Dacko et al. (1996) suggest that force feedback is distributed according to muscle articulation rather than fiber type because slow-twitch motor units in the MG were not inhibited.

Additional support for this hypothesis was obtained using data from the quadriceps muscles (Wilmink, 1998). In this muscle group, the variables of articulation and fiber type could be separated. The vastus muscles were found to be linked almost exclusively by strong excitatory feedback from muscle spindles and not by inhibitory force feedback (figure 9.1, b and c) in spite of the uniformly slow-twitch motor units in vastus intermedius. Force feedback was found between each vastus muscle and the biarticular rectus femoris (RF) muscle (figure 9.1c) (Wilmink, 1998). In addition, the excitatory length feedback was particularly weak between RF and the vastus muscles (Eccles et al., 1957a; Wilmink, 1998). The distributions of both forms of feedback therefore supported the hypothesis that articulation was the major determinant of the distributions. The force feedback and weak excitatory feedback between RF and the vastus muscles are also consistent with the functional dissociation of these muscles during locomotion. The vastus muscles provide knee extension moments during stance, whereas RF becomes active during both stance and swing in proportions determined by the speed of locomotion (Engberg and Lundberg, 1969). These findings indicate that inhibitory force feedback mediates coordination across joints in parallel with the direct mechanical linkages of biarticular muscles.

Inhibitory force feedback was found to link muscles across axes of rotation in addition to articular joints (figure 9.1, a and c). For example, members of the triceps surae group are linked to the peroneus muscles (abductors) as well as to the tibialis posterior muscle (adductor) by force-related inhibition (Bonasera and Nichols, 1994, 1996). If muscle groups are defined by muscles with common actions and mutual excitation from muscle spindle receptors, then force feedback links members of different muscle groups (Nichols, 1994). Furthermore, these neural pathways are organized in parallel with the mechanical linkages of biaxial muscles (Nichols, 1994; Perot and Goubel, 1982). For example, the axes of abduction/adduction and plantarflexion/dorsiflexion are mechanically linked in the cat hindlimb by the triceps surae muscles and also by neural feedback between the triceps surae muscles and the peroneus brevis muscle. In summary, the major feature that characterizes the distribution of inhibitory force feedback is that muscles are linked across different joints and different axes of rotation. This organization suggests that these pathways serve to coordinate the limb according to its major degrees of freedom. As background tension increases, the coupling provided by force feedback also increases, while the coupling due to the combination of intrinsic properties and autogenic reflexes of biarticular muscles tends to remain constant (as noted earlier). Therefore, the mechanical coupling of joints increases with exertion.

Inhibitory force feedback across joints, the mechanical linkages of biarticular muscles, and the effects of length feedback on these biarticular muscles (figure 9.1) all tend to compensate for the effects of nonuniform distribution of segment inertia, as described earlier. Forcible dorsiflexion of the ankle would evoke reflex responses from the group Ia mediated pathways linking muscles crossing the ankle. The mechanical effect of the gastrocnemius muscles on the knee and the inhibitory neural feedback from the triceps surae to the quadriceps muscles (Wilmink, 1998) would result in a flexion moment of the knee and a decrease in the extensor moment of the ankle, respectively. These two mechanisms would mediate the distribution of the perturbation to the more proximal joints of the limb and the reduction of the large apparent impedance of the more proximal limb segments. Therefore, these neural and mechanical influences promote interjoint coordination and stability of the limb.

Another property of the musculoskeletal system that can potentially lead to instability is the reduction in moment arms of muscles at the extremes of joint motion (Lawrence and Nichols, 1999a; Lieber and Boakes, 1988). Such changes in moment arm could lead to instability if torque production were to decline as well. A reduction of mechanical coupling between joints could also result from a decline in moment arm at one or both joints spanned by a given muscle. Since force feedback depends upon muscle force rather than torque, the ability of this feedback system to contribute to interjoint coordination would be independent of changes in moment arm as long as the muscles receiving the force feedback were able to generate significant torque. For example, if the gastrocnemius muscles were to lose the ability to generate flexor torque about an extended knee, the forces generated could still provoke substantial force-dependent inhibition onto the quadriceps muscles.

The organization of inhibitory force feedback contrasts with the distribution of two other feedback systems. First, there are excitatory, polysynaptic pathways that arise from both muscle spindles and Golgi tendon organs. This "group I excitation" is expressed during locomotion (Guertin et al., 1995; Pearson, 1995) and apparently is widely distributed among antigravity muscles (Guertin et al., 1995). The function of this pathway is apparently to mediate a "loading reflex" that enhances antigravity action in the face of changing loads (Dietz et al., 1992). During quiet standing, both excitatory and inhibitory effects have been observed in response to intramuscular stimulation (Pratt, 1995), an adequate stimulus for Golgi tendon organs, suggesting the coexistence of positive and negative force feedback. Second, an additional set of pathways arising from group III and group IV receptors that collectively mediate clasp-knife inhibition is also distributed across antigravity muscles (Cleland and Rymer, 1990; Cleland et al., 1990), regardless of the joints spanned by these muscles. However, these inhibitory effects act somewhat more strongly on single-joint muscles, suggesting that coordination in the limb is preserved by the action of multiarticular muscles during the resulting flexion (Nichols and Cope, 2001). The distribution and presumed function of these two feedback systems contrast with the unique attributes of the short-latency inhibition from Golgi tendon organs in providing a neural linkage between joints and axes of rotation. Since the expression of neural systems mediating "group I excitation" and clasp-knife inhibition appears to be task-specific, the short-latency, inhibitory pathways may be more generally applicable to the regulation of joint mechanics and interjoint coordination.

Conclusion

It may be provisionally concluded from the foregoing discussion that short-latency, length-related excitation and force-related inhibition strongly influence the me-chanical properties of limbs. Length feedback increases muscular stiffness and decreases the dependence of stiffness on the amplitude of the disturbance and on background force (stiffness regulation). The system of length feedback that links synergistic and antagonistic muscles regulates the stiffness of the joint in three dimensions (Benati et al., 1980). The strength of neural feedback for the muscles that pull in a transverse plane at the ankle is particularly strong to provide the required stabilization. By virtue of biarticular and biaxial muscles, length feedback enhances force transmission between joints and axes of rotation and therefore pro-motes limb stability. The efficacy of this force transmission as well as the directional stiffness properties of the joint are not strongly affected by activation levels of the musculature due to stiffness regulation by length feedback.

Force feedback from Golgi tendon organs increases with background force and is more prolonged in action than length feedback. Force feedback spans joints and axes of rotation and parallels the mechanical linkages of biarticular and biaxial muscles. Inhibitory force feedback, mechanical coupling across axes and joints, and length feedback to biarticular and biaxial muscles act synergistically to promote coordination between joints and rotation axes. For example, a disproportionately large perturbation delivered to the ankle joint will result in a compensatory reduc-tion in stiffness of the knee due to these three mechanisms. These include the mechanical linkages of the gastrocnemius muscles that produce a flexion moment, the enhancements of the stiffnesses of these muscles due to stretch reflexes, and inhibitory force feedback from the triceps surae to the quadriceps muscles that reduces the extension moment at the knee. In the time domain, inhibitory force feedback develops somewhat more slowly than length feedback (Nichols, 1999) and may outlast it, providing a time-dependent degree of interjoint coupling.

The dependence of the strength of force feedback on background force suggests that the degree of interjoint coupling is also state-dependent. At low forces, where force feedback is small, the mechanical properties of the limb should be dominated by mechanical linkages and length feedback. Since intrinsic stiffness is low under these conditions, length feedback accounts for a large portion of the coupling. At higher forces, the immediate response of the limb to perturbations is still dominated by intrinsic properties and length feedback, but length feedback accounts for a smaller proportion of the stiffness. As force feedback increases with time, compliance of the limb increases and joint motions become more tightly coupled, promoting coordination. As proposed in the original version of the stiffness hypothesis, length and force feedback act in opposition to regulate limb stiffness. In contrast, these pathways act together to promote interjoint coordination.

References

Abelew, T.A., Miller, M.D., Cope, T.C., and Nichols, T.R. (2000) Local loss of proprioception results in disruption of interjoint coordination during locomotion in the cat. *J Neurophysiol* 84: 2709-2719.

Benati, M., Gaglio, S., Morasso, P., Tagliasco, V., and Zaccaria, R. (1980) Anthropomorphic robotics: II. Analysis of manipulator dynamics and the output motor impedance. *Biol Cybern* 38: 141-150.

Bonasera, S.J., and Nichols, T.R. (1994) Mechanical actions of heterogenic reflexes linking long toe flexors and extensors of the knee and ankle in the cat. *J Neurophysiol* 71: 1096-1110.

Bonasera, S.J., and Nichols, T.R. (1996) Mechanical actions of heterogenic reflexes among ankle stabilizers and their interactions with plantarflexors of the cat hindlimb. *J Neurophysiol* 75(5): 2050-2070.

Burkholder, T.J., and Nichols, T.R. (2000). The mechanical action of proprioceptive length feedback in a model of the cat hindlimb. *Motor Control* 4: 201-220.

Cleland, C.L., Hayward, L., and Rymer, R.W. (1990) Neural mechanisms underlying the clasp-knife reflex in the cat: II. Stretch-sensitive muscular-free nerve endings. *J Neurophysiol* 64: 1319-1330.

Cleland, C.L., and Rymer, W.Z. (1990) Neural mechanisms underlying the clasp-knife reflex in the cat: I. Characteristics of the reflex. *J Neurophysiol* 64: 1303-1318.

Cope, T.C., Bonasera, S.J., and Nichols, T.R. (1994) Reinnervated muscles fail to produce stretch reflexes. *J Neurophysiol* 71: 817-820.

Cope, T.C., and Clark, B.D. (1993) Motor-unit recruitment in self-reinnervated muscle. *J Neurophysiol* 70: 1787-1796.

Dacko, S.M., Sokoloff, A.J., and Cope, T.C. (1996) Recruitment of triceps surae motor units in the decerebrate cat: I. Independence of type S units in soleus and medial gastrocnemius muscles. *J Neurophysiol* 75(5): 1997-2004.

Dietz, V., Gollhofer, A., Kleiber, M., and Trippel, M. (1992) Regulation of bipedal stance: Dependency on 'load' receptors. *Exp Brain Res* 89: 229-231.

Eccles, J.C., Eccles, R.M., and Lundberg, A. (1957a) The convergence of monosynaptic excitatory afferents onto many different species of alpha motoneurons. *J Physiol* 137: 22-50.

Eccles, J.C., Eccles, R.M., and Lundberg, A. (1957b) Synaptic actions on motoneurons caused by impulses in the Golgi tendon organ afferents. *J Physiol* 138: 227-252.

Eccles, R.M., and Lundberg, A. (1958) Integrative pattern of Ia synaptic actions on motoneurons of hip and knee muscles. *J Physiol* 144: 271-298.

Edgley, S., Jankowska, E., and McCrea, D. (1986) The heteronymous monosynaptic actions of triceps surae group Ia afferents on hip and knee extensor motoneurones in the cat. *Exp Brain Res* 61: 443-446.

Engberg, I., and Lundberg, A. (1969) An electromyographic analysis of muscular activity in the hindlimb of the cat during unrestrained locomotion. *Acta Physiol Scand* 75: 614-630.

Goslow, G.E., Reinking, R.M., and Stuart, D.G. (1973) The cat step cycle: Hindlimb joint angles and muscle lengths during unrestrained locomotion. *J Morphol* 141: 1-42.

Guertin, P., Angel, M., Perreault, M.-C., and McCrea, D.A. (1995) Ankle extensor group I afferents excite extensors throughout the hindlimb during MLR-evoked fictive locomotion in the cat. *J Physiol* 487: 197-209.

Hayward, L., Breitbach, D., and Rymer, W.Z. (1988) Increased inhibitory effects on close synergists during muscle fatigue in the decerebrate cat. *Brain Res* 440: 199-203.

Hoffer, J.A., and Andreassen, S. (1981) Regulation of soleus muscle stiffness in premammillary cat intrinsic and reflex components. *J Neurophysiol* 45: 267-285.

Hogan, N. (1985) The mechanics of multi-joint posture and movement control. *Biol Cybern* 52: 315-331.

Hollister, A., Buford, W.L., Myers, L.M., Giurintano, D.J., and Novick, A. (1992) The axes of rotation of the thumb carpometacarpal joint. *J Orthop Res* 10: 454-460.

Houk, J.C. (1972) The phylogeny of muscular control configurations. In Drischel, H., and Dettmer, P. (Eds.), *Biocybernetics,* vol. 4, pp. 125-144. Jena, Germany: Fischer.

Houk, J.C. (1979) Regulation of stiffness by skeletomotor reflexes. *Ann Rev Physiol* 41: 99-114.

Houk, J.C., Crago, P.E., and Rymer, W.Z. (1981a) Function of the spindle dynamic response in stiffness regulation: A predictive mechanism provided by non-linear feedback. In Taylor, A., and Prochazka, A. (Eds.), *Muscle receptors and movement,* pp. 299-309. London: Macmillan.

Houk, J.C., Harris, D.A., and Hasan, Z. (1973) Non-linear behavior of spindle receptors. In Stein, R.B., Pearson, K.G., Smith, R.S., and Redford, J.B. (Eds.), *Advances in behavioral biology. Control of posture and locomotion,* vol. 7, pp. 147-163. New York: Plenum.

Houk, J.C., and Henneman, E. (1974) Feedback control of muscle: Introductory concepts. In Mountcastle (Ed.), *Medical physiology* (13th ed.), vol. 1, pp. 608-616. St. Louis: C.V. Mosby.

Houk, J.C., and Rymer, W.Z. (1981) Neural control of muscle length and tension. In Brooks, V.B. (Ed.), *Handbook of physiology,* The nervous system, Vol. II. Motor control, Part 1, pp. 257-324. Bethesda: American Physiological Society.

Houk, J.C., Rymer, W.Z., and Crago, P.E. (1981b) Dependence of dynamic response of spindle receptors on muscle length and velocity. *J Neurophysiol* 46: 143-165.

Houk, J.C., Singer, J.J., and Goldman, M.R. (1970) An evaluation of length and force feedback to soleus muscles of decerebrate cats. *J Neurophysiol* 33: 784-811.

Hoy, M.G., and Zernicke, R.F. (1985) Modulation of limb dynamics in the swing phase of locomotion. *J Biomech* 18: 49-60.

Hultborn, H. (1976) Transmission in the pathway of reciprocal Ia inhibition to motoneurones and its control during the tonic stretch reflex. In Homma, S. (Ed.), *Understanding the stretch reflex,* pp. 235-255. Amsterdam: Elsevier.

Huyghues-Despointes, C.M.J.I. (1998) *Effects of movement history on the intrinsic properties and the neural regulation of feline skeletal muscle.* Emory University, Atlanta.

Kirsch, R.F., and Rymer, W.Z. (1992) Neural compensation for fatigue-induced changes in muscle stiffness during perturbations of elbow angle in human. *J Neurophysiol* 68: 449-470.

Laporte, Y., and Lloyd, D.P.C. (1952) Nature and significance of the reflex connections established by large afferent fibers of muscular origin. *Am J Physiol* 169: 609-621.

Lawrence, J.H.I., and Nichols, T.R. (1999a) A three-dimensional biomechanical analysis of the cat ankle joint complex: II. Effects of ankle joint orientation on evoked isometric joint torques. *J Appl Biomech* 15: 106-119.

Lawrence, J.H.I., and Nichols, T.R. (1999b) A three-dimensional biomechanical analysis of the cat ankle joint complex: I. Active and passive postural mechanisms. *J Appl Biomech* 15: 95-105.

Lawrence, J.H.I., Nichols, T.R., and English, A.W. (1993) Cat hindlimb muscles exert substantial torques outside the sagittal plane. *J Neurophysiol* 69: 282-285.

Lieber, R.L., and Boakes, J.L. (1988) Sarcomere length and joint kinematics during torque production in frog hindlimb. *Am J Physiol* 254: 759-768.

Lin, D.C., and Rymer, W.Z. (1993) Mechanical properties of cat soleus muscle elicited by sequential ramp stretches: Implications for control of muscle. *J Neurophysiol* 70: 997-1008.

Lloyd, D.P.C. (1946) Integrative pattern of excitation and inhibition in two-neuron reflex arcs. *J Neurophysiol* 9: 439-444.

Macpherson, J.M. (1988a) Strategies that simplify the control of quadrupedal stance: I. Forces at the ground. *J Neurophysiol* 6: 204-217.

Macpherson, J.M. (1988b) Strategies that simplify the control of quadrupedal stance: II. Electromyographic activity. *J Neurophysiol* 60: 218-231.

Macpherson, J.M. (1994) The force constraint strategy for stance is independent of prior experience. *Exp Brain Res* 101: 397-405.

McCrea, D.A. (1986) Spinal cord circuitry and motor reflexes. *Exerc Sport Sci Rev* 14: 105-141.

Mussa-Ivaldi, F.A., Hogan, N., and Bizzi, E. (1985) Neural, mechanical, and geometric factors subserving arm posture in humans. *J Neurosci* 5(10): 2732-2743.

Nichols, T.R. (1989) The organization of heterogenic reflexes among muscles crossing the ankle joint in the decerebrate cat. *J Physiol* 410: 463-477.

Nichols, T.R. (1994) A biomechanical perspective on spinal mechanisms of coordinated muscular action: An architecture principle. *Acta Anat* 151: 1-13.

Nichols, T.R. (1999) Receptor mechanisms underlying heterogenic reflexes among the triceps surae muscles of the cat. *J Neurophysiol* 81: 467-478.

Nichols, T.R., and Cope, T.C. (2001) The organization of distributed proprioceptive feedback in the chronic spinal cat. In Cope, T.C. (Ed.), *Motor neurobiology of the spinal cord*, pp. 305-326. Boca Raton: CRC Press.

Nichols, T.R., Cope, T.C., and Abelew, T.A. (1999) Rapid spinal mechanisms of motor coordination. *Exerc Sport Sci Rev* 27: 255-284.

Nichols, T.R., and Houk, J.C. (1976) The improvement in linearity and regulation of stiffness that results from action of the stretch reflex. *J Neurophysiol* 39: 119-142.

Nichols, T.R., Lin, D.C., and Huyghues-Despointes, C.M.J.I. (1999) The role of musculoskeletal mechanics in motor coordination. In Binder, M.D. (Ed.), *Progress in brain research*, pp. 369-378. Amsterdam: Elsevier.

Pearson, K.G. (1995) Proprioceptive regulation of locomotion. *Curr Opin Neurobiol* 5: 786-791.

Perot, C., and Goubel, F. (1982) Synergy between bifunctional muscles at the ankle joint. *Eur J Appl Physiol* 48: 59-65.

Pratt, C.A. (1995) Evidence of positive force feedback among hindlimb extensors in the intact standing cat. *J Neurophysiol* 73: 2578-2583.

Prochazka, A. (1996) Proprioceptive feedback and movement regulation. New York: Oxford.

Proske, U., Morgan, D.L., and Gregory, J.E. (1992) Muscle history dependence of responses to stretch of primary and secondary endings of cat soleus muscle spindles. *J Physiol* 445: 81-95.

Rymer, W.Z., and Hasan, Z. (1980) Absence of force-feedback regulation in soleus of the decerebrate cat. *Brain Res* 184: 203-209.

Thomson, D.B., Inglis, J.T., Schor, R.H., and Macpherson, J.M. (1991) Bilateral labyrinthectomy in the cat: Motor behaviour and quiet stance parameters. *Exp Brain Res* 85: 364-372.

Watt, D.G.D., Stauffer, E.K., Taylor, A., Reinking, R.M., and Stuart, D.G. (1976) Analysis of muscle receptor connections by spike-triggered averaging: 1. Spindle primary and tendon organ afferents. *J Neurophysiol* 39: 1375-1378.

Wilmink, R.J.H. (1998) *Organization and modulation of excitatory and inhibitory spinal reflexes in cats and humans.* Aalborg University, Aalborg.

Zytnicki, D., Lafleur, J., Horcholle-Bossavit, G., Lamy, F., and Jami, L. (1990) Reduction of Ib autogenetic inhibition in motoneurons during contractions of an ankle extensor muscle in the cat. *J Neurophysiol* 64: 1380-1389.

10

Steadiness
of Lengthening Contractions

Evangelos A. Christou, Brian L. Tracy, and Roger M. Enoka

Department of Kinesiology and Applied Physiology,
University of Colorado at Boulder

When a muscle is used to displace a load, the change in muscle length depends on the relative magnitudes of the muscle torque and the opposing load torque about the joint. When the muscle and load torques are equal, for example, the net torque about the joint is zero and the muscle performs an isometric contraction. When the magnitude of the muscle torque is greater than the load torque, the load will be lifted as the muscle performs a shortening contraction. In contrast, when the magnitude of the load torque is greater than the muscle torque, the load will be decreased as the muscle performs a lengthening contraction. If the task is to decrease a load with a prescribed trajectory, the muscle torque must be controlled precisely so that it is slightly less than the load torque throughout the entire movement.

Research on differences between shortening and lengthening contractions has focused on the mechanical properties of muscle and on the adaptive capabilities of the neuromuscular system (Enoka, 1996). The classic force-velocity relation of muscle, for example, indicates that the maximum force a muscle can exert at a given contraction velocity is different for shortening and lengthening contractions (Flitney and Hirst, 1978; Hill, 1938; Katz, 1939) and that EMG amplitude is lower during lengthening contractions at comparable forces (Bigland and Lippold, 1954).

Similarly, the strength gain achieved after several weeks of physical training is specific to the type of contraction performed in the training program (Higbie et al., 1996; Hortobágyi et al., 1996; Hortobágyi et al., 1997). Furthermore, the use of lengthening contractions appears to result in greater muscle damage and soreness compared with other types of contractions (Chleboun et al., 1998; Malm et al., 1999; McHugh et al., 1999; Proske and Morgan, 2001; Whitehead et al., 1998).

Accumulating evidence suggests that the control strategies used by the nervous system differ among the contraction types. Such findings include subjects' inability to maximally activate a muscle during a voluntary lengthening contraction (Aagaard et al., 2000; Webber and Kriellaars, 1997; Westing et al., 1990) and differences in the magnitude of evoked potentials between shortening and lengthening contractions (Abbruzzese et al., 1994; Nardone and Schieppati, 1988; Sekiguchi et al., 2001). Furthermore, there is a greater resistance to fatigue during voluntary, but not during artificially, evoked lengthening contractions (Binder-Macleod and Lee, 1996; Tesch et al., 1990), greater synchronization of motor unit discharge during lengthening contractions (Semmler et al., 2001), and increased cortical potentials as measured by electroencephalography during lengthening contractions (Fang et al., 2001). The most controversial observations, however, have been differences in the recruitment order of motor units during lengthening compared with shortening contractions (Howell et al., 1995; Nakazawa et al., 1993; Nardone et al., 1989; Nardone and Schieppati, 1988; c.f. Bawa and Jones, 1999; Christova and Kossev, 2000; Kossev and Christova, 1998; Sogaard et al., 1996).

Because the discharge behavior of motor units can influence one's ability to perform a steady contraction (Laidlaw et al., 2000; Taylor et al., 2000), we have compared the steadiness of shortening and lengthening contractions. Steadiness is measured as the fluctuations of force or acceleration in the time domain but can also be expressed in the frequency domain; selected bandwidths within the frequency domain are quantified as tremor (McAuley and Marsden, 2000). The functional significance of steadiness is that it influences a person's ability to exert a precise force and to perform consistent movement trajectories. Although the steadiness of isometric and shortening contractions has been studied for more than a century (Fullerton and Cattell, 1892), the steadiness of lengthening contractions has received much less attention. The purpose of this chapter is to examine factors that contribute to differences in steadiness between shortening and lengthening contractions.

Lengthening Contractions Are Less Steady Within and Across Trials

Several experimental tasks have been used to compare steadiness of shortening and lengthening contractions (figure 10.1). These tasks include matching velocity or force templates, movements at various speeds, and movements with different types of loads (e.g., inertial, isokinetic).

The findings of experiments with inertial loads indicate that when subjects attempt to match a constant-velocity template with light loads, the performance is less

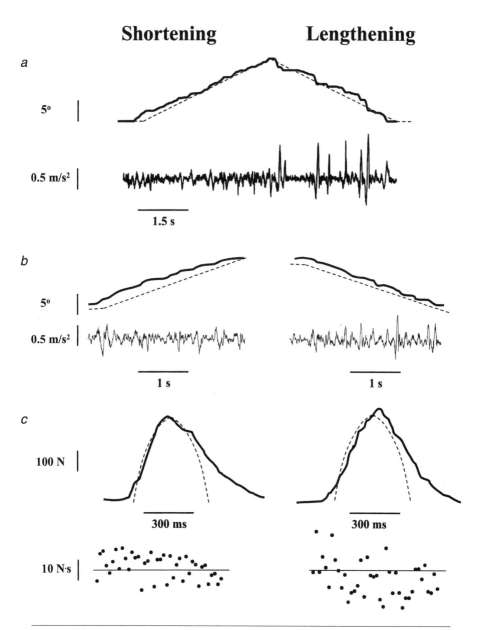

Shortening   **Lengthening**

a

5°

0.5 m/s²

1.5 s

b

5°

0.5 m/s²

1 s   1 s

c

100 N

300 ms   300 ms

10 N·s

Figure 10.1   Representative data for shortening and lengthening contractions as performed in three different tasks. *(a, b)* Show position (top trace) and acceleration (bottom trace) during shortening and lengthening contractions performed in *(a)* a fixed or *(b)* a randomized order. *(c)* Shows representative force data (top) from a single trial in which a subject attempted to match a force-time parabola during repetitive rapid contractions on an isokinetic dynamometer. The variability of impulse (force-time integral) around the mean for 40 separate contractions is shown at the bottom of *(c)*.

steady during lengthening contractions compared with shortening contractions (Burnett et al., 2000; Graves et al., 2000; Laidlaw et al., 2000; Tracy et al., in press). In these studies, the shortening contractions typically preceded the lengthening contractions, and the movement speed was slow (<0.5 rad/s). When the order of the contractions was reversed, however, no differences in steadiness were found between shortening and lengthening contractions (Laidlaw et al., 2001). To eliminate the potential order effect and examine the influence of movement speed on steadiness, we compared the steadiness of shortening and lengthening contractions that were performed at various speeds in a random order. Each contraction was performed separately and was always preceded by an isometric contraction (Christou et al., 2001). We found that the steadiness of a contraction decreased with movement speed and that lengthening contractions produced greater fluctuations in acceleration compared with shortening contractions. Furthermore, the declines in steadiness of movement speed were greater for the lengthening contractions compared with shortening contractions.

The focus of these studies was the fluctuations of acceleration within each trial. However, motor output also varies across trials. Trial-to-trial variability is often assessed from rapid contractions (~200 ms) that are performed repetitively, in which the subjects attempt to match a prescribed force-time parabola (Carlton and Newell, 1993; Christou and Carlton, 2001). Two such studies have examined trial-to-trial variability of the knee extensor muscles during shortening and lengthening contractions performed on an isokinetic dynamometer (Christou and Carlton, 1999; Christou and Carlton, in press). The first study examined the trial-to-trial variability of shortening and lengthening contractions performed by young individuals over a range of absolute target forces (50-250 N; N = newton), whereas the second study compared the performance of young and old individuals to relative target forces (20-90% of the force achieved during a maximum voluntary contraction [MVC]). The two studies found that the trial-to-trial variability in peak force, force-time integral, and temporal characteristics of the movement was greater during lengthening contractions compared with shortening contractions (figure 10.1c). A similar result was observed when subjects performed constant-velocity contractions while lifting and lowering an inertial load with the first dorsal interosseus muscle (Christou et al., 2001).

With the exception of one study, current evidence indicates that lengthening contractions are not only less steady than shortening contractions within a trial but also are more variable across trials. Although these findings are consistent across various experimental tasks, the differences between shortening and lengthening contractions appear to be greater with light loads and during rapid tasks.

Old Adults Are Less Steady During Lengthening Contractions

Several studies indicate that the ability of old adults (> 65 years of age) to perform a steady isometric contraction is impaired (Burnett et al., 2000; Keen et al., 1994;

Laidlaw et al., 2000; Laidlaw et al., 1999; Semmler et al., 2000b; Tracy and Enoka, 2002). Some of these studies examined the effect of age on the steadiness of shortening and lengthening contractions. Findings from these studies suggest that the ability of old adults to perform steady lengthening contractions declines to a greater extent than for shortening contractions. For example, old adults exhibit greater fluctuations in displacement (Laidlaw et al., 2000) and acceleration (Burnett et al., 2000) of the index finger than young adults do when lowering light inertial loads with the first dorsal interosseus muscle (figure 10.2a). Similarly, fluctuations in acceleration during lowering of light loads with the elbow flexor muscles were greater for old adults compared with young adults, especially during lengthening contractions (Graves et al., 2000) (figure 10.2b).

Not all studies, however, have found that the steadiness of lengthening contractions is reduced in old adults. For example, similar fluctuations were observed in acceleration and displacement for the elbow flexor (Tracy et al., in press) and knee extensor (Tracy and Enoka, 2002; Tracy et al., in press) muscles of young and old adults. Although old adults produce less steady movements while lifting and lowering light loads (<15% maximum) with the first dorsal interosseus and elbow flexor muscles (Burnett et al., 2000; Graves et al., 2000; Laidlaw et al., 2000), old adults are at least as steady as young adults when lifting heavier loads (Burnett et al., 2000; Graves et al., 2000) (figure 10.2). Furthermore, when the lengthening contractions preceded shortening contractions, steadiness was similar for the young and old subjects (Laidlaw et al., 2001). In addition, when active old adults are compared with young adults at various movement speeds during separate lengthening and shortening contractions that are preceded by an isometric contraction, old adults experience fewer fluctuations in acceleration (Christou et al., 2001). Nonetheless, the relative increase in the standard deviation of acceleration is greater for old adults during lengthening contractions. These findings indicate that differences in the steadiness of lengthening contractions between young and old adults vary with the muscle group used, the type of task performed, and the speed of the movement.

When old adults perform the same contraction repetitively, however, their performance is consistently more variable, especially during lengthening contractions. For example, when active old adults perform constant-velocity contractions while lifting or lowering loads with the first dorsal interosseus muscle, they exhibit greater trial-to-trial variability compared with young adults during lengthening contractions (Christou et al., 2001). Similarly, old adults demonstrate greater trial-to-trial variability than young adults when they reproduce a force-time parabola (20-90% MVC) with the knee extensor muscles (Christou and Carlton, in press). This variability, which is most evident during lengthening contractions, is the result of an impaired ability to reproduce the timing characteristics of the contraction (figure 10.3).

Findings from the various studies presented suggest that the ability of old adults to perform steady lengthening contractions declines only with light loads. The variability in force from trial to trial, furthermore, is impaired, especially during lengthening contractions. Such factors as physical activity, experimental task, and the load lifted influence the steadiness of old adults' lengthening contractions.

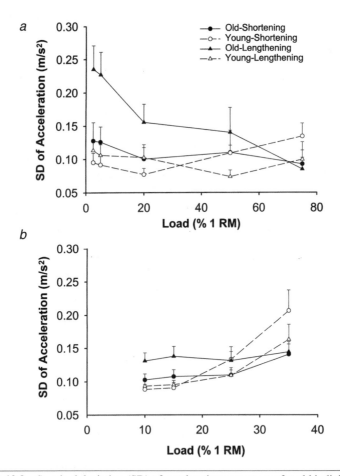

Figure 10.2   Standard deviation (SD) of acceleration was greater for old individuals during slow lengthening contractions with light loads for *(a)* the first dorsal interosseus and *(b)* elbow flexor muscles.

Steadiness of Lengthening Contractions Varies Among Muscle Groups

A series of studies in different muscle groups suggests that the difference in steadiness between shortening and lengthening contractions depends on the muscle group studied. For example, fluctuations in displacement during contractions of the first dorsal interosseus muscle were greater during lowering of loads compared with lifting of loads that ranged from 2.5 to 10% of maximum (Laidlaw et al., 2000). A similar finding was observed for the fluctuations in acceleration during lowering of loads with the first dorsal interosseus muscle (Burnett et al., 2000). In these two studies, the differences in steadiness between shortening and lengthening contrac-

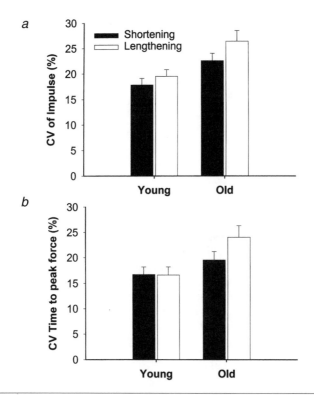

Figure 10.3   Between-trial variability (coefficient of variation, CV) for *(a)* the force-time integral (impulse) and *(b)* the time to peak force during shortening and lengthening contractions with the knee extensor muscles. Old adults exhibited greater variability for impulse and time to peak force during lengthening contractions.

tions were attributable to the greater fluctuations during lengthening contractions in old adults compared with young adults. Nevertheless, subsequent findings suggest that the lengthening contractions with the first dorsal interosseus also are less steady in young adults compared with their shortening contractions (Christou et al., 2001; Semmler et al., 2001). Comparable results were also found during lifting and lowering of loads (10 and 15% of maximum) with the elbow flexor muscles (Graves et al., 2000). In contrast, fluctuations in displacement were similar during shortening and lengthening contractions of the knee extensor muscles with loads that ranged from 5 to 50% of maximum (Tracy and Enoka, 2002).

Although the experimental design and methods were similar in these separate studies on hand, arm, and leg muscles, the studies used different subject cohorts. Accordingly, subtle differences in the characteristics of the subject samples might partially explain differences observed among the studies. To minimize this confounding influence, we have measured fluctuations in acceleration during shortening and lengthening contractions of the first dorsal interosseus, elbow flexor, and

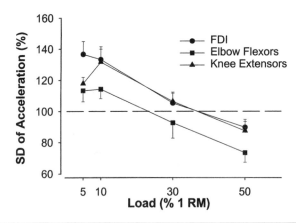

Figure 10.4 Standard deviation (SD) of acceleration during lengthening contractions expressed as a percentage of the value obtained for the shortening contractions performed with the first dorsal interosseu (FDI), elbow flexor, and knee extensor muscles. Values greater than 100% indicate a less steady performance during the lengthening contractions. The difference between lengthening and shortening contractions appears to be greater for the first dorsal interosseus and knee extensor muscles compared with the elbow flexor muscles. For all three muscles, lengthening contractions appear to be less steady with light loads and steadier with heavy loads compared with shortening contractions.

knee extensor muscles in the same individuals (Tracy et al., in press). A preliminary analysis of these data provides further evidence of differences among muscle groups.

Lengthening contractions with light loads (5-10% of maximum) were less steady compared with shortening contractions for the first dorsal interosseus and the knee extensor muscles but not with the elbow flexor muscles (Tracy et al., in press) (figure 10.4). Furthermore, it appears that the first dorsal interosseus exhibits a greater difference between shortening and lengthening contractions with the lightest load (5% of maximum) compared with the elbow flexor and knee extensor muscles (figure 10.4).

Taken together, these findings suggest that differences in steadiness between shortening and lengthening contractions vary among muscle groups. The differences among studies, therefore, appear to be confounded by differences in experimental design and the muscle examined.

Neural Mechanisms

Several features of the activation signal generated by the nervous system can influence the steadiness of lengthening contractions. These include details of the motor output from the spinal cord, the organization of the descending command, and the integration of feedback from sensory receptors.

Recent interest in the neural control of lengthening contractions was heightened by the observation (Nardone et al., 1989) that the recruitment order of motor units may deviate from that prescribed by the size principle (Henneman, 1957). Nardone et al. (1989) found that high-threshold motor units were selectively activated during lengthening contractions with the triceps surae muscles. Furthermore, when motor unit activity was recorded from the first dorsal interosseus during lifting and lowering of a load, Howell et al. (1995) found that 3 high-threshold motor units out of 21 recorded were selectively recruited during lengthening contractions. In contrast, several studies have failed to find evidence of selective activation of high-threshold motor units during lengthening contractions (Bawa and Jones, 1999; Christova and Kossev, 2000; Kossev and Christova, 1998; Laidlaw et al., 2000; Søgaard et al., 1996). These results suggest that although recruitment order may be disrupted during lengthening contractions, the most common strategy is not to alter the recruitment order (Enoka and Fuglevand, 2001).

Nonetheless, muscle activity can shift from slow to fast muscles during lengthening contractions. Such findings were observed in the triceps surae muscles when shortening contractions involved greater activation of the soleus muscle (80% type I muscle fibers) and lengthening contractions involved greater activation of the gastrocnemius muscle (50% type II muscle fibers) (Nardone and Schieppati, 1988). Selective activation of muscles during lengthening contractions has also been observed during movements produced by the elbow flexor muscles (Nakazawa et al., 1993). The relative activation of the brachioradialis muscle changed through the range of motion during lengthening contractions but not during shortening contractions. Altered distribution among synergist muscles during lengthening contractions, therefore, may potentially influence the steadiness of a movement.

The most consistent difference in motor output between shortening and lengthening contractions is the amplitude of muscle activation. Studies have found that the EMG amplitude is significantly lower during lengthening contractions compared with shortening contractions with similar muscle torques (Bawa and Jones, 1999; Burnett et al., 2000; Christou et al., 2001; Howell et al., 1995; Kossev and Christova, 1998; Laidlaw et al., 2000; Nardone and Schieppati, 1988; Sogaard et al., 1996). The differences in EMG activity during voluntary movements increased with movement speed (Christou et al., 2001; Kossev and Christova, 1998; Westing et al., 1991). Similarly, reduced levels of activation are observed in experiments where motor evoked potentials and H-reflexes are elicited during shortening and lengthening contractions (Abbruzzese et al., 1994; Nardone et al., 1989; Sekiguchi et al., 2001). This reduction in EMG amplitude during lengthening contractions is partially due to a decrease in discharge rate of the motor units (Kossev and Christova, 1998; Laidlaw et al., 2000; Søgaard et al., 1996). For example, motor units discharge at lower frequencies during lengthening contractions with low loads (<10% of maximum) in the first dorsal interosseus and biceps brachii muscles (Howell et al., 1995; Sogaard et al., 1996) and with moderate loads (50% of maximum) in the triceps brachii (Kossev and Christova, 1998). At lower discharge rates, the twitches of individual motor units are less fused and cause greater fluctuations in muscle force during low-force contractions (Fuglevand et al., 1993). Furthermore, the discharge rate of motor units is more variable during eccentric contractions (Laidlaw et al., 2000), which computer simulations indicate can cause greater fluctuations in force (Taylor et al., 2000).

The discharge of action potentials by motor units may also become more correlated during lengthening contractions. The summation of twitch forces that result from this synchronous discharge of motor units can potentially contribute to the observed differences in steadiness between shortening and lengthening contractions (Semmler, 2002). For example, when the level of motor unit synchronization was varied while all other factors were kept constant, the steadiness of a computer-simulated contraction declined with increases in the level of synchronous discharge (Yao et al., 2000). Experimental evidence also provides support for this finding. Synchronous discharge was quantified in pairs of motor units during isometric, shortening, and lengthening contractions with the first dorsal interosseus muscle (Semmler et al., 2001; Semmler et al., 2000a). Lengthening contractions were less steady and exhibited higher levels of synchronization compared with shortening contractions. Furthermore, the synchronization of motor unit discharge was correlated with the steadiness of isometric and anisometric contractions.

Because motor unit synchronization is an index of the strength of common input to motor neurons (Semmler, 2002), the descending command may differ during lengthening contractions. Several studies support this possibility. For example, responses evoked in the biceps brachii by magnetic and electrical stimulation of the motor cortex were reduced during lowering compared with lifting of a load (Abbruzzese et al., 1994). Furthermore, differences in motor-evoked potentials between shortening and lengthening contractions have been observed during various intensities of electrical stimulation of the motor cortex (Sekiguchi et al., 2001). Specifically, the plateau value and maximum slope of the sigmoidal curve that characterizes the relation of evoked potential size with intensity of stimulation were lower for lengthening contractions. In addition, movement-related cortical potentials, as measured by electroencephalography, were greater during lengthening contractions compared with shortening contractions (Fang et al., 2001). These findings indicate that a constant excitatory input to the motor cortex during the two types of contractions results in a reduced output for lengthening contractions, whereas when the output is the same between the two contractions the cortical input is greater for lengthening contractions.

In addition to differences in the motor output from the spinal cord and the descending command, lengthening contractions appear to involve qualitative differences in the integration of sensory feedback. For example, data from primates (Schieber and Thach, 1985) and humans (Burke et al., 1978) show that muscle spindle afferents discharge at a higher rate and for a longer duration during lengthening contractions. Furthermore, the sign of the short-latency response to a cutaneous stimulus is reversed during lengthening contractions compared with shortening contractions (Haridas et al., 2000). One potential explanation for these differences in reflex responses is that the role of sensory feedback is more critical during lengthening compared with shortening contractions. Experimental evidence suggests that when the contribution of sensory feedback to the control of a movement is limited, such as occurs in rapid movements, there is a further reduction in the steadiness of lengthening contractions (Christou et al., 2001).

Summary

The motor output from the spinal cord to muscle differs during lengthening compared with shortening contractions. These differences, which include the recruitment of fewer motor units, a lower and more variable discharge rate, and a greater amount of motor unit synchronization, likely contribute to the differences in steadiness between shortening and lengthening contractions. Variation in the motor output is a consequence of differences in the net input received by the motor neurons, which are due to changes in both the descending drive and the integration of sensory feedback.

References

Aagaard, P., Simonsen, E.B., Andersen, J.L., Magnusson, S.P., Halkjaer-Kristensen, J., and Dyhre-Poulsen, P. (2000). Neural inhibition during maximal eccentric and concentric quadriceps contraction: Effects of resistance training. *J Appl Physiol* 89: 2249-2257.

Abbruzzese, G., Morena, M., Spadavecchia, L., and Schieppati, M., (1994). Response of arm flexor muscles to magnetic and electrical brain stimulation during shortening and lengthening tasks in man. *J Physiol* 481: 499-507.

Bawa, P., and Jones, K.E. (1999). Do lengthening contractions represent a case of reversal in recruitment order? *Prog Brain Res* 123: 215-220.

Bigland, B., and Lippold, O.C.J. (1954). The relation between force, velocity, and integrated electrical activity in human muscles. *J Physiol* 123: 214-224.

Binder-Macleod, S.A., and Lee, S.C. (1996). Catchlike property of human muscle during isovelocity movements. *J Appl Physiol* 80: 2051-2059.

Burke, D., Hagbarth, K.E., and Lofstedt, L. (1978). Muscle spindle activity in man during shortening and lengthening contractions. *J Physiol* 277: 131-142.

Burnett, R.A., Laidlaw, D.H., and Enoka, R.M. (2000). Coactivation of the antagonist muscle does not covary with steadiness in old adults. *J Appl Physiol* 89: 61-71.

Carlton, L.G., and Newell, K.M. (1993). Force variability and characteristics of force production. In: *Force variability*, edited by Newell K.M. and Cordo P. Champaign, IL: Human Kinetics, p. 128-132.

Chleboun, G.S., Howell, J.N., Conatser, R.R., and Giesey, J.J. (1998). Relationship between muscle swelling and stiffness after eccentric exercise. *Med Sci Sports Exerc* 30: 529-535.

Christou, E.A., and Carlton, L.G. (1999). Motor output variability during concentric and eccentric contractions of the quadriceps femoris muscle group. *Proceedings of the 23rd Annual Meeting of the American Society of Biomechanics*, Pittsburg, PA, p. 129-130.

Christou, E.A., and Carlton, L.G. (2001). Old adults exhibit greater motor output variability than young adults only during rapid discrete isometric contractions. *J Gerontol A Biol Sci Med Sci* 56: B524-B532.

Christou, E.A., and Carlton, L.G. (in press). Age and contraction type influence motor output variability in rapid discrete tasks. *J Appl Physiol*.

Christou, E.A., Shinohara, M., and Enoka, R.M. (2001). The changes in EMG and steadiness with variation in movement speed differ for concentric and eccentric contractions. *Proceedings of the 25th Annual Meeting of the American Society of Biomechanics*, San Diego, CA., p. 333-334.

Christova, P., and Kossev, A. (2000). Human motor unit activity during concentric and eccentric movements. *Electromyogr Clin Neurophysiol* 40: 331-338.

Enoka, R.M. (1996). Eccentric contractions require unique activation strategies by the nervous system. *J Appl Physiol* 81: 2339-2346.

Enoka, R.M., and Fuglevand, A.J. (2001). Motor unit physiology: Some unresolved issues. *Muscle Nerve* 24: 4-17.

Fang, Y., Siemionow, V., Sahgal, V., Xiong, F., and Yue, G.H. (2001). Greater movement-related cortical potential during human eccentric versus concentric muscle contractions. *J Neurophysiol* 86: 1764-1772.

Flitney, F.W., and Hirst, D.G. (1978). Cross-bridge detachment and sarcomere "give" during stretch of active frog's muscle. *J Physiol* 276: 449-465.

Fuglevand, A.J., Winter, D.A., and Patla, A.E. (1993). Models of recruitment and rate coding organization in motor-unit pools. *J Neurophysiol* 70: 2470-2488.

Fullerton, G.S., and Cattell, J.M. (1892). On the perception of small differences. (philosophical monograph series no. 2). Philadelphia, PA: University of Pennsylvania Press.

Graves, A.E., Kornatz, K.W., and Enoka, R.M. (2000). Older adults use a unique strategy to lift inertial loads with the elbow flexor muscles. *J Neurophysiol* 83: 2030-2039.

Haridas, C., Zehr E.P., Sugajima, Y., and Gillies, E. (2000). Differential control of cutaneous reflexes during lengthening and shortening contractions of the human triceps surae. *Society for Neuroscience Abstracts* 26: 1232.

Henneman, E. (1957). Relation between size of neurons and their susceptibility to discharge. *Science* 126: 1345-1347.

Higbie, E.J., Cureton, K.J., Warren, G.L. III, and Prior, B.M. (1996). Effects of concentric and eccentric training on muscle strength, cross- sectional area, and neural activation. *J Appl Physiol* 81: 2173-2181.

Hill, A.V. (1938). The heat of shortening and the dynamic constants of muscle. *Proc Royal Soc London* B 126: 136-195.

Hortobágyi, T., Hill, J.P., Houmard, J.A., Fraser, D.D., Lambert, N.J., and Israel, R.G. (1996). Adaptive responses to muscle lengthening and shortening in humans. *J Appl Physiol* 80: 765-772.

Hortobágyi, T., Lambert, N.J., and Hill, J.P. (1997). Greater cross education following training with muscle lengthening than shortening. *Med Sci Sports Exerc* 29: 107-112.

Howell, J.N., Fuglevand, A.J., Walsh, M.L., and Bigland-Ritchie, B. (1995). Motor unit activity during isometric and concentric-eccentric contractions of the human first dorsal interosseus muscle. *J Neurophysiol* 74: 901-904.

Katz, B. (1939). The relation between force and speed in muscular contraction. *J Physiol* 96: 45-64.

Keen, D.A., Yue, G.H., and Enoka, R.M. (1994). Training-related enhancement in the control of motor output in elderly humans. *J Appl Physiol* 77: 2648-2658.

Kossev, A., and Christova, P. (1998). Discharge pattern of human motor units during dynamic concentric and eccentric contractions. *Electroencephalogr Clin Neurophysiol* 109: 245-255.

Laidlaw, D.H., Bilodeau, M., and Enoka, R.M. (2000). Steadiness is reduced and motor unit discharge is more variable in old adults. *Muscle Nerve* 23: 600-612.

Laidlaw, D.H., Hunter S.K., and Enoka, R.M. (2001). Slow eccentric contractions are not always less steady than concentric contractions for old adults. *Proceedings of the 25th Annual Meeting of the American Society of Biomechanics*, San Diego, CA., p. 311-312.

Laidlaw, D.H., Kornatz, K.W., Keen, D.A., Suzuki, S., and Enoka, R.M. (1999). Strength training improves the steadiness of slow lengthening contractions performed by old adults. *J Appl Physiol* 87: 1786-1795.

Malm, C., Lenkei, R., and Sjodin, B. (1999). Effects of eccentric exercise on the immune system in men. *J Appl Physiol* 86: 461-468.

McAuley, J.H., and Marsden, C.D. (2000). Physiological and pathological tremors and rhythmic central motor control. *Brain* 123: 1545-1567.

McHugh, M.P., Connolly, D.A., Eston, R.G., and Gleim, G.W. (1999). Exercise-induced muscle damage and potential mechanisms for the repeated bout effect. *Sports Med* 27: 157-170.

Nakazawa, K., Kawakami, Y., Fukunaga, T., Yano, H., and Miyashita, M. (1993). Differences in activation patterns in elbow flexor muscles during isometric, concentric and eccentric contractions. *Eur J Appl Physiol* 66: 214-220.

Nardone, A., Romanò, C., and Schieppati, M. (1989). Selective recruitment of high-threshold human motor units during voluntary isotonic lengthening of active muscles. *J Physiol* 409: 451-471.

Nardone, A., and Schieppati, M. (1988). Shift of activity from slow to fast muscle during voluntary lengthening contractions of the triceps surae muscles in humans. *J Physiol* 395: 363-381.

Proske, U., and Morgan, D.L. (2001). Muscle damage from eccentric exercise: mechanism, mechanical signs, adaptation and clinical application. *J Physiol* 537: 333-334.

Schieber, M.H., and Thach, W.T., Jr. (1985). Trained slow tracking II. Bidirectional discharge patterns of cerebellar nuclear, motor cortex, and spindle afferent neurons. *J Neurophysiol* 54: 1228-1270.

Sekiguchi, H., Kimura, T., Yamanaka, K., and Nakazawa, K. (2001). Lower excitability of the corticospinal tract to transcranial magnetic stimulation during lengthening contractions in human elbow flexors. *Neurosci Lett* 312: 83-86.

Semmler, J.G. (2002). Motor unit synchronization and neuromuscular performance. *Exerc Sports Sci Rev* 30: 8-14.

Semmler, J.G., Kornatz, K.W., Kern, D.S., and Enoka, R.M. (2001). Motor unit synchronization reduces the steadiness of anisometric contractions by a hand muscle. [CD-ROM; Program No. 168.3]. *Society for Neuroscience Abstracts* 27.

Semmler, J.G., Kutzscher, D.V., Zhou, S., and Enoka, R.M. (2000a). Motor unit synchronization is enhanced during slow shortening and lengthening contractions of the first dorsal interosseus muscle. *Society for Neuroscience Abstracts* 26: 463.

Semmler, J.G., Steege, J.W., Kornatz, K.W., and Enoka, R.M. (2000b). Motor-unit synchronization is not responsible for larger motor-unit forces in old adults. *J Neurophysiol* 84: 358-366.

Søgaard, K., Christensen, H., Jensen, B.R., Finsen, L., and Sjøgaard, G. (1996). Motor control and kinetics during low level concentric and eccentric contractions in man. *Electroencephalogr Clin Neurophysiol* 101: 453-460.

Taylor, A., Steege, J., and Enoka, R.M (2000). Increased variability of motor unit discharge rate decreases the steadiness of simulated isometric contractions. *The Physiologist* 43: 321.

Tesch, P.A., Dudley, G.A., Duvoisin, M.R., Hather, B.M., and Harris, R.T. (1990). Force and EMG signal patterns during repeated bouts of concentric or eccentric muscle actions. *Acta Physiol Scand* 138: 263-271.

Tracy, B.L., and Enoka, R.M. (2002). Older adults are less steady during submaximal isometric contractions with the knee extensor muscles. *J Appl Physiol* 92: 1004-1012.

Tracy, B.L., Mehoudar, P.D., Ortega, J.D., and Enoka, R.M. (in press). The steadiness of isometric contractions is similar between upper and lower extremity muscle groups. *Med Sci Sports Exerc Suppl.*

Webber, S., and Kriellaars, D. (1997). Neuromuscular factors contributing to in vivo eccentric moment generation. *J Appl Physiol* 83: 40-45.

Westing, S.H., Cresswell, A.G., and Thorstensson, A. (1991). Muscle activation during maximal voluntary eccentric and concentric knee extension. *Eur J Appl Physiol Occup Physiol* 62: 104-108.

Westing, S.H., Seger, J.Y., and Thorstensson, A. (1990). Effects of electrical stimulation on eccentric and concentric torque-velocity relationships during knee extension in man. *Acta Physiol Scand* 140: 17-22.

Whitehead, N.P., Allen, T.J., Morgan, D.L., and Proske, U. (1998). Damage to human muscle from eccentric exercise after training with concentric exercise. *J Physiol* 512: 615-620.

Yao, W., Fuglevand, R.J., and Enoka, R.M. (2000). Motor-unit synchronization increases EMG amplitude and decreases force steadiness of simulated contractions. *J Neurophysiol* 83: 441-452.

11

On the Problem
of Adequate Language
in Biology

Israel M. Gelfand
Rutgers University

Mark L. Latash
Penn State University

*We would like to dedicate this chapter to the memory
of two outstanding scientists,
Nikolai Alexandrovich Bernstein and Mikhail L'vovich Tsetlin.*

The current volume contains chapters that discuss different aspects of control of biological movement. Each chapter typically focuses on one particular level of analysis of the complex system for movement generation and its possible changes with neurological disorder, development, and practice. Combining such papers within one volume has been a major purpose of this series of publications and corresponding conferences "Progress in Motor Control." However, one should be frank: We are still far from having a common language that would link all these studies together. Very basic questions remain without an answer. Among them are "What is controlled during a movement?" and "What is the difference between biological movement and movement of an inanimate object?" These questions are poorly formulated, and this is the main reason they are so difficult to address.

The purpose of the current chapter is to develop a framework, a set of notions that would allow us to formulate these general questions in a more exact way. The recent progress in the development of ideas expressed by the Russian school of motor control about 40 years ago makes us optimistic in our ability to formulate such questions in a way that allows their direct experimental testing. We believe that such a framework is applicable to different biological phenomena, not only to motor behavior.

Adequate Language

Having an adequate language (adequate set of notions and ideas) is a necessary prerequisite for development of any area of science. For example, the early period of the development of geometry led to the formulation of an adequate language for analysis of spatial relations among objects. Such a language allowed discussion of geometry problems in terms of reasoning. An example is the definition of a point as an object that does not have width, length, and height. This definition has no logical meaning and has been reconsidered in the 20th century, but it was a major element in the development of an adequate language for geometry.

Successful scientific analysis of movements of inanimate objects has been based on an adequate language reflected in the apparatus of differential equations. However, for biological objects, the main problem is not in the description of their mechanics or other physicochemical properties. Even the simplest movement of lifting the arm, which can be described well with forces and coordinates, turns into a science fiction novel if the process is considered within its biological context. Thus, for a biological system, an adequate language has to be more transcendental, at least from a mathematician's viewpoint.

Recent studies within biological subdisciplines have commonly focused on specific problems formulated using notions and ideas imported directly from such fields as control theory, certain areas of physics, and more recently, from the theory of nonlinear differential equations (for reviews, see Feldman and Levin, 1995; Kelso, 1995; Latash, 1993; Stein, 1982). These studies have led to numerous important findings and interesting conclusions. The problem, however, is that these studies have tried to squeeze all phenomena into the Procrustean bed of available mathematical theories rather than to take a step backward, to the origins of the problem, to look at it without any mathematical or other preconceptions and then move forward.

The urgent need for an adequate language (or languages) that would make biology a science of its own has become obvious (cf. Gelfand and Latash, 1998). We believe that we are presently at an early stage of the development of an adequate language(s) for biological sciences.

Elements of History

N.A. Bernstein (1935, 1947, 1967) was probably the first to realize that specific features of behavior of biological objects cannot be reduced to specific features of elements that make up these objects. Bernstein further suggested that the development of the central nervous system had been driven by the design of the peripheral motor apparatus (i.e., its numerous mechanical degrees of freedom that needed to be controlled). He concluded that the central nervous system was built on a hierarchical principle.

Ideas of Bernstein were developed by Gelfand and Tsetlin (1962, 1966), who introduced the principle of nonindividualized control, which preserves the idea of hierarchical control but rejects prescriptive, authoritarian control by hierarchically higher levels. According to the principle of nonindividualized control, elements of a complex system are not controlled individually but are united into task-specific or intention-specific *structural units*. Structural units are organized in a flexible, task-specific way for purposes that can be termed *synergies*. External behavior generated by a structural unit will be defined by its purpose (i.e., by the corresponding synergy) and by current external conditions.

Structural units can be introduced for systems of different complexity. For example, a cell, a subsystem within the human body, an organism, or a group of organisms can each be viewed as a structural unit. Gelfand and Tsetlin (1966) introduced three major axioms to describe the essential properties of structural units:

1. The internal structure of a structural unit is always more complex than its interaction with the environment (which may include other structural units).

2. Part of a structural unit cannot itself be a structural unit with respect to the same group of tasks.

3. Parts of a structural unit that do not work with respect to a task either

   a. are eliminated or

   b. find their own places within the task.

Axiom 3a illustrates the principle of economy when a minimal number of elements carry out each given task. Hence, the set of axioms 1, 2, and 3a is more applicable to description of established, stereotyped reactions.

Axiom 3b illustrates the very important *principle of abundance,* later developed by Gelfand (1991), where many more elements than necessary participate in the activity of a structural unit with respect to each task. The set of axioms 1, 2, and 3b describes systems that may be expected to evolve and form structural units able to solve tasks for which it had not originally been designed. The principle of abundance is considered typical of movement organization in higher animals.

Now we can say that an adequate language operates with structural units. A structural unit of a system is characterized by the fact that the number of internal connections within a structural unit is at least an order of magnitude higher than the number of its external connections. A structural unit is not simply an assembly of elements (e.g., neurons) but a system with a particular function. Changes of connections among the elements may lead to the creation of different structural units based on the same set of elements.

Illustration of a Structural Unit

The notion of structural unit and the three major axioms can be illustrated using, as an example, the organization of a scientific laboratory. The laboratory can be viewed as a structural unit, while individual researchers are its elements. Let us assume that the laboratory has been organized to solve a specific problem (e.g., to design a new running shoe). The ultimate result of the functioning of the laboratory (the shoe) can be described much more simply than the way researchers interact with each other (axiom 1). If the laboratory functions properly, half of it cannot solve the same problem (axiom 2).

The laboratory may be organized based on axiom 3a or on axiom 3b. In the first case, a minimal number of researchers are hired, and each researcher is assigned a unique, specific function. In this case, the laboratory will be able to fulfill its purpose successfully, but it will be unable to perform any other task. It will also be unable to perform the original task if one of the members of the laboratory suddenly falls ill or retires.

In the second case, a large group of talented researchers is assembled and asked to deal with the problem. Each researcher is expected to find his or her own place in the team and contribute to the process. Such a team will be able to solve the original task but can also be expected to reorganize successfully if the task is modified (e.g., if it involves designing a new leg prosthesis). Such a laboratory may also be expected to continue functioning successfully if one of the members quits.

Problem of Redundancy or Solution of Abundance?

The problem of motor redundancy was originally introduced by Bernstein (1935, 1947, 1967) as one of finding a unique solution when more elements than absolutely necessary participate in a particular motor task. Bernstein illustrated this problem with the task of pointing to a target in the three-dimensional space with the tip of the index finger. The number of independent axes of joint rotation within the human arm allows an infinite number of solutions for this problem. In contemporary literature, such problems are addressed as *problems of inverse kinematics* (Mussa-Ivaldi et al., 1989). If one wants to define patterns of individual joint torques that would ensure a certain trajectory of the endpoint, another Bernstein-type problem is encountered, commonly termed the *problem of inverse dynamics* (Atkeson, 1989).

Similar problems may be formulated at other levels of the human system for movement production (Latash, 1996). For example, which forces should be produced by each of the six major muscles crossing the elbow joint to generate a certain value of elbow joint torque (cf. Prilutsky, 2000), or which motor units should be recruited by the central nervous system and at which frequencies to produce a certain level of muscle activation? Solving such ill-posed problems is equivalent to solving an equation with several unknowns. Apparently, these problems do not have unique solutions unless additional constraints are introduced into the task formulation.

Many attempts have been made to address the problem of motor redundancy in its original formulation. These involved, in particular, application of optimization principles based on certain mechanical, engineering, psychological, or complex cost functions (for reviews, see Latash, 1993; Seif-Naraghi and Winters, 1990). However, we would like to suggest that the problem itself has been inadequately formulated and needs to be reconsidered.

The main reason for this suggestion is a ubiquitous characteristic of voluntary movements, i.e., motor variability. Studies of motor variability have turned into a large area of their own (for a review, see Newell and Corcos, 1993). However, probably the most important observation was made in the twenties by Bernstein (1923) in his famous studies of blacksmiths hitting the chisel with the hammer. These were the best-trained subjects one could imagine because they had performed the same labor movement hundreds of times a day for years.

Bernstein used a very sophisticated method of movement analysis (kimocyclography) that he had developed himself (Bernstein, 1927). Within this method, small lightbulbs were placed at certain important points on the subject's body, and the movements were photographed on a slowly moving film using a high-speed lens shutter. This produced a series of snapshots of the bulbs, allowing Bernstein to calculate the motion of individual joints and of particular points on the subject's body. In his studies, Bernstein reached the frequency of more than 500 snapshots per second (Bernstein and Popova, 1929), which is comparable to the most sophisti-cated optoelectronic systems that are used for movement analysis today.

In the study of blacksmiths, Bernstein noticed that the variability of the tip of the hammer across a series of trials was smaller than the variability of individual joints of the subject's right arm holding the hammer. He concluded that the joints were not acting independently but were correcting each other's errors. This observation sug-gests that the central nervous system does not try to find a unique solution for the problem of kinematic redundancy but uses the apparently redundant set of joints to ensure more accurate (less variable) performance of the task.

The study of the blacksmiths illustrates the general idea expressed earlier (axiom 3b) as the principle of abundance. One may conclude that the principle of abundance renders the problem of redundancy irrelevant: Numerous elements should not be viewed as a source of computational problems for the nervous system but as a useful apparatus that requires proper organization.

A particular method for dealing with abundant systems, the method of ravines, was introduced by Gelfand and Tsetlin (1966). This method was later modified to explain certain observations of the wiping reflex in the decerebrate frog (Berkinblit et al., 1986a, 1986b).

We hope that readers will forgive the two authors, neither of whom is a native English speaker, for the following linguistic suggestion: The word *redundancy* has a negative connotation as something of no use that needs to be eliminated, whereas *abundance* has a positive connotation as something that may be used rather than eliminated. In the original writings, Bernstein used the Russian word *izbytochnost,* which originates from the noun *izbytok* and can mean "redundancy" or "abundance," depending on the context. An imprecise translation of this single word has led to the development of a whole direction of research commonly addressed as the "redundancy problem" or the "Bernstein problem."

The Principle of Minimal Interaction

Within a hierarchical organization, a functional goal is formulated by an upper level in the hierarchy, but this formulation does not suppress the freedom of elements at the lower level of the hierarchy. In the 1960s, Gelfand and Tsetlin (1966) suggested a principle of minimal interaction, which states that the interaction among elements at the lower level of the hierarchy is organized so as to minimize the external input to each of the elements. This principle means that elements interact so that each of them tries to keep its preferred mode of action while the functional output is simultaneously kept at a desired level and does not require intervention from the higher level of the hierarchy, even when one or more of the elements change their contributions.

In a way, elements of a structural unit may be compared to a class of lazy students whose main purpose is to keep the teacher from giving them new assignments (figure 11.1). When a task is specified by the teacher, the students interact in such a way that their overall output (e.g., the level of noise in the classroom, shown by the dashed arrow in figure 11.1) keeps the teacher happy or, even better, asleep. If the class becomes too quiet or too noisy, the teacher is likely to wake up and give a new task.

The principle of minimal interaction implies, in particular, that if a perturbation is applied to one of the elements of a structural unit, it is expected to lead to changes in the contributions of other elements, not only of the perturbed one. The purpose of these changes is to correct errors in the common functional output of the structural unit that were introduced by the changed contribution of the perturbed element. Quick corrections to unexpected perturbations during natural, complex movements

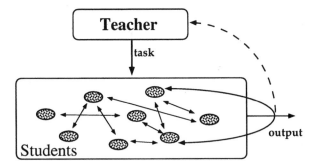

Figure 11.1 A schematic representation of a structural unit. The hierarchically higher supreme controller (teacher) specifies only a few important task parameters (task), while interactions among the elements (students) create an output. The teacher will interfere only when the feedback on the output (the broken line) is outside an acceptable range. The students try to maintain the feedback within the range so that no new assignment is given.

have been reported in a number of effectors for such different activities as speech production (Abbs and Gracco, 1984; Saltzman and Munhall, 1989), arm motor tasks (Gielen et al., 1988; Latash, 2000), vertical posture maintenance (Cordo and Nashner, 1982; Nashner, 1976; Nashner and Woollacott, 1979), and locomotion (Dietz et al., 1984; Forssberg, 1979; Grillner, 1979).

Examples From the Area of Motor Behavior

Let us illustrate the notion of structural unit with a few examples. We have purposefully selected these examples to represent a broad range of motor studies, from a spinal reflex in a frog to the coordinated action of the human joints, fingers, and limbs. These examples show how the described approach can bring together such seemingly different phenomena in very different species.

Wiping Reflex in the Spinal Frog

If a small piece of paper soaked in a weak acid solution is placed on the back or a forelimb of a frog whose spinal cord has been surgically cut at a high level, the frog wipes off the stimulus with a coordinated multijoint motion of the ipsilateral hindlimb (the wiping reflex). The wiping movement remains accurate in conditions of different position of the forelimb with the stimulus (Fookson et al., 1980) and in conditions of hindlimb loading and joint fixation (Berkinblit, Feldman, Fookson, Latash, unpublished; cited in Latash, 1993). A model has been suggested by Berkinblit et al. (1986b) within which the vector of angular velocity in each joint is defined by a relation between two vectors, one from the joint to the target and the other from the joint to the limb endpoint (figure 11.2). This model has an inherent ability to generate accurate movements even if one of the joints gets off track because of an unexpected loading or joint fixation (i.e., it has a built-in error intercompensation among the joints).

Note that a more recent model by Rosenbaum and his colleagues (Rosenbaum et al., 1995) also demonstrates a feature of error intercompensation based on a different computational approach.

These examples illustrate the principle of minimal interaction since the interaction among signals to individual joints is organized in such a way that it tends to eliminate the effects of changes in the output of one of the joints on the important functional outcome. As a result, a hierarchically higher level (a "controller") does not need to interfere in cases of perturbations that can be dealt with by the structural unit itself.

Finger Coordination

The main principles of organization of structural units have been actively studied in the field of control of voluntary movements. A recent series of studies of finger coordination during force production by a group at Penn State University (Danion et al., 2000; Latash et al., 1998a, 1998b; Li et al., 1998a, 1998b, 2000; Zatsiorsky et al., 1998, 2000) may be viewed as an example.

In these studies, it has been shown that forces of individual fingers are not controlled independently so that the fingers of a hand form a structural unit (cf. the principle of nonindividualized control). The following indices of multifinger coordination have been described:

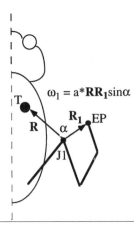

$$\omega_1 = a*RR_1 \sin\alpha$$

Figure 11.2 During the wiping reflex, the spinal frog wipes an irritating stimulus away from its back with a movement of the ipsilateral hindlimb (shown schematically as a kinematic chain of rigid links). The angular velocity of each joint is assumed to represent a product of two vectors: from the joint to the endpoint of the limb and from the joint to the target. This model can generate successful wiping even if individual joint motion is perturbed.

1. *Sharing:* Total force is shared among the fingers in a certain fashion over a wide range of force values (Li et al., 1998a). The sharing pattern (a combination of percentages of total force generated by each finger) is stable under informational perturbations (Latash et al., 1998b) and under fatigue of all four fingers (Danion et al., 2000, 2001). The sharing pattern can change if external mechanical conditions are changed, in particular, when fingers are involved in a gripping task and the position of the opposing thumb varies (Li et al., 1998b).

2. *Deficit:* When several fingers act together and try to produce maximal total force, peak forces of individual fingers are smaller than during single-finger maximal voluntary contraction (MVC) tasks (Li et al., 1998a). In young subjects, the average magnitude of force deficit per finger during four-finger tasks is about 10% to 20% of corresponding MVCs.

3. *Enslaving:* When a subject is instructed to produce force with only one, two, or three fingers (master fingers), explicitly noninvolved fingers (slave fingers) also generate forces (Li et al., 1998a). These forces may be very high, reaching over 50% of the MVC of slave fingers. The enslaving effects are nonadditive (i.e., effects from one master finger may be higher than effects from two master fingers acting simultaneously) (Zatsiorsky et al., 1998, 2000).

A principle of minimization of secondary moments has been suggested as an organizing principle during multifinger force production tasks (Li et al., 1998a). This principle states that forces are shared among fingers in such a way that the total moment created by all the fingers with respect to the longitudinal axis of the hand is minimized. This principle has been shown to be valid in a variety of tasks. An analysis of finger enslaving patterns (Zatsiorsky et al., 2000) has also suggested that they are organized so that the secondary moment is reduced. The principle of minimization of secondary moments has been shown to be task-specific. Its violations have been reported for tasks with self-imposed perturbations (Latash et al., 1998a).

In one study (Li et al., 1998a), subjects were asked to produce ramp forces with four fingers acting in parallel from 0 to MVC over 5 s. Variances of the peak total force and of peak individual finger forces were then compared across trials. Note that if the fingers act as independent force generators, the sum of the variances of individual finger forces [Var (F_i), i = 1, 2, 3, 4] should be equal to the variance of the total force [Var (F_{tot})]:

$$\text{Var } (F_{tot}) = \text{Var } (F_1) + \text{Var } (F_2) + \text{Var } (F_3) + \text{Var } (F_4) \qquad (11.1)$$

If, however, fingers compensate for each other's errors, the following inequality should take place:

$$\text{Var } (F_{tot}) < \text{Var } (F_1) + \text{Var } (F_2) + \text{Var } (F_3) + \text{Var } (F_4) \qquad (11.2)$$

In other words, if by chance one finger in one trial develops a higher force than usual, one or more of the other fingers will likely produce a smaller force than usual. All 12 subjects who took part in this experiment showed a relation among force variances that fitted inequality (11.2) but not equation (11.1). Hence, the hypothesis on error compensation has been corroborated, illustrating the principle of minimal interaction.

In another study (Latash et al., 1998a), the subjects were asked to produce a moderate constant level of force (25% of MVC) with the index, middle, and ring fingers acting in parallel. They were then asked to tap with one of the fingers at a frequency of 1 Hz, specified by a metronome. Apparently, the tapping finger decreased its contribution to the total force when it lost contact with the surface. At the initiation of tapping, a simultaneous increase in the forces generated by the nontapping fingers was observed (figure 11.3). Cross-correlation functions between individual finger forces showed peaks at close to zero time delay, suggesting a feedforward mechanism of finger force adjustment. This increase in the forces by nontapping fingers during the first cycle of tapping compensated for 94% to 102% of the total force loss due to the lifting of the finger that performed the tapping. This finding can also be viewed as an example of error compensation among the fingers if one views maintaining constant force as a major component of the task and tapping as a self-inflicted perturbation.

Bimanual Coordination

When a waiter lifts a heavy pitcher off a tray supported by the other hand, motor commands to muscles controlling the supporting hand change in parallel with commands to the hand that lifts the pitcher (Dufosse et al., 1985). Otherwise, an upward

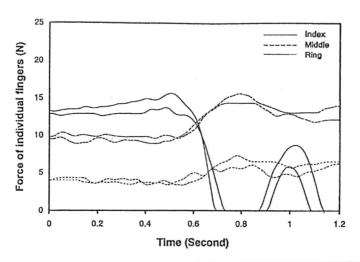

Figure 11.3   The subject maintained a constant level of force with three fingers and then performed a few taps with the index finger. Two trials are illustrated. Note the increase in the force of the other fingers simultaneous with the beginning of the tapping.

Reprinted from Latash, Li, and Zatsiorsky 1998.

motion of the tray would occur, and potentially the tray could be dropped. Thus, the two hands are united into a structural unit and obey the principle of minimal interaction.

We have recently extended this observation to a different task in which one hand grasps a cylindrical object and the other hand provides supporting force by pressing against the bottom of the object (Scholz and Latash, 1998). In this series, subjects sat with their dominant forearm supported up to the wrist while holding a cylindrical "cup" between their thumb and fingers (an opposition grip). Force transducers recorded the grip force applied normally to the cup side by the thumb and to the cup bottom. On different series, a supporting force was added to and released from the bottom of the cup by the subject's nondominant hand or by the experimenter (figure 11.4a). We also recorded electromyograms (EMGs) of a number of hand and forearm muscles. Within this task, the mechanical conditions did not require changes in the contribution of the grasping hand when a supporting force was applied; however, the application of the bottom force was accompanied by a feedforward drop in the grip force; such effects were not seen when the supporting force was applied by the experimenter (figure 11.4b). This drop in the grip force can be viewed as controlled by a factor related to keeping the safety margin constant (cf. Johansson and Westling, 1984). The results suggest that grip force adjustments and bottom force changes represented peripheral patterns of a single central process (cf. the principle of minimal interaction) rather than being separately controlled focal and postural components of the action.

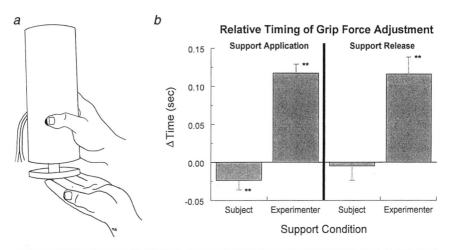

Figure 11.4   *(a)* The subject gripped a "cup" with the fingers of one hand and applied supporting force to the bottom of the cup with the other hand. *(b)* Time delays between the force applied to or released from the bottom and changes in the grip force averaged across subjects. Grip force changes start earlier (negative values) when the bottom force is applied by the subject; grip force changes start at a delay of about 100 ms when the bottom force is applied by the experimenter.

Reprinted from Scholz and Latash 1998.

The Uncontrolled Manifold Hypothesis

A new, quantitative approach to the problem of coordination of many elements, directly related to the principle of minimal interaction, has recently been suggested by Schöner and Scholz (Scholz and Schöner, 1999; Schöner, 1995). According to this approach, setting a functional task may be associated with selecting a performance variable that is selectively stabilized with respect to possible perturbations (internal and external). In other words, a structural unit is created with an interaction among its elements organized in such a way that the selected performance variable shows less variance across different realizations of the task than other performance variables.

If one considers a time sequence of states of a multielement system, this process may be viewed as selecting a manifold within the state space of elements for each instant of time. This manifold is such that changes in elements' states within the manifold do not change a certain desired instantaneous value of the selected performance variable. After such a manifold is selected, individual elements are allowed to change their states as long as they remain within the manifold but not if they leave the manifold. One may say that the elements are less controlled within the manifold than outside it; hence, the manifold has been termed *uncontrolled manifold* (UCM) and the hypothesis has been designated as the "UCM-hypothesis."

For example, imagine that a person is asked to generate a constant force by pressing with two hands against a stationary object. There are two ways of making sure that the total force shows low variability. First, the forces of each hand may be controlled rather precisely so that they both show low variability and, as a result, their sum shows low variability (data sets D1, D2, and D3 in figure 11.5a). Alternatively, a relation between the hand forces may be established such that an increase in the force of one hand is accurately matched with a decrease in the force of the other hand (data set D4 in figure 11.5b). A line in the two-dimensional space of hand forces, corresponding to this relation, is an uncontrolled manifold for this particular task. If a number of measurements are performed for such a task, individual points on the force-force plane may show more or less circular clouds (figure 11.5a) or an ellipsis-shaped cloud with the long axis of the ellipsis coinciding with the uncontrolled manifold (figure 11.5b).

A priori, one could suggest that a multielement system could be organized without selecting an uncontrolled manifold but simply by stabilizing a state of the system equally in all directions within the state space. Within the mental experiment with two-hand force production described earlier, the subject could maintain low total force variability by maintaining the forces of both hands nearly constant rather than allowing them to covary (if the clouds of points in figure 11.5a are small). The UCM-hypothesis takes one additional step and suggests that this is not how control of voluntary movements is organized. This illustration shows that the UCM-hypothesis is not trivial and that control can be organized without selecting a UCM.

The principle of minimal interaction implies that corrections are introduced into states of individual elements when one or more elements change their states (are

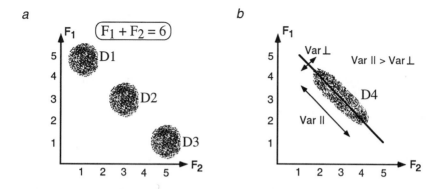

Figure 11.5   An illustration of the uncontrolled manifold concept. The subject is asked to produce a force of 6 units by pressing with two hands. The task can be performed by limiting the variability of each force (as in clouds of data points D1, D2, and D3 in *[a]*) or by establishing a relation between the two forces (an uncontrolled manifold illustrated by the straight line in *[b]*) that keeps the total force at the required level. In the latter case, the cloud of data points (D4) is expected to show larger variance along the UCM than perpendicular to the UCM.

perturbed) so that successful execution of a functionally important task is jeopardized. If one associates high stability of a selected performance variable with successful execution of a task, the UCM-hypothesis allows finding a particular solution within the principle of minimal interaction.

Testing Control Hypotheses With the UCM Method

An important feature of an uncontrolled manifold is that it is specific to a particular performance variable that is presumed to be selectively stabilized by the subject's central nervous system. That is why an analysis of a particular movement using the UCM-hypothesis needs to be based on a "control hypothesis" (i.e., a hypothesis as to what performance variable is important within a particular task). The word *control* is used in this context as preferential stabilization of a variable as compared to all other performance variables. In this sense, a number of variables may be controlled within each task, with each variable being controlled to a greater or lesser degree (reflected in a more or less elongated ellipsoid of data points similar to the D4 set illustrated in figure 11.5b).

The first step in performing such an analysis is to formulate a "control hypothesis" (i.e., to select performance variables that may be preferentially stabilized during the task). In the earlier example of force production with two hands, the control hypothesis was that the subject tried to produce constant total force. After a variable is selected, a manifold may be calculated within the state space of the elements of the whole system that preserves a particular value of this variable corresponding to successful task performance. Then, over a number of trials, an

index of variability of element states (variance) is decomposed into two components: one that lies within the uncontrolled manifold and one that is perpendicular
to the manifold (figure 11.5).

An assessment of components of joint variability within the uncontrolled manifold
and perpendicular to the manifold allows generation of a quantitative index of selective
stabilization of the variable for which the manifold has been calculated. For example,
the ratio of the two components can be calculated and used as such an index.

One recently published study used the UCM-hypothesis to analyze the movement of standing up from a sitting position (Scholz and Schöner, 1999). In this study,
the authors looked at changes in joint angles across a number of standing-up movements by subjects. They then formulated several control hypotheses, calculated
corresponding uncontrolled manifolds, and analyzed joint variance as consisting of
the previously mentioned two components. Their results have demonstrated, in
particular, that the position of the body's center of mass in a sagittal plane was a
controlled variable. The horizontal head position was controlled to a lesser degree.
The vertical position of the head was not controlled (i.e., its variability was not
selectively reduced by an interaction among the joints).

Joint Interaction During Pistol Shooting

The UCM-hypothesis has been recently applied to an analysis of multijoint coordination involved in a quick-draw shooting task using an infrared pistol (Scholz,
Schöner, and Latash, 2000). Note that accurate shooting depends crucially on two
angles describing the pistol orientation with respect to the target. However, it is
rather independent of the actual position of the pistol and of the angle of its rotation
about the axis of the barrel. These lay observations suggest that an uncontrolled
manifold in the seven-dimensional space of the major arm joint angles may be
formed to ensure accurate shooting. A corresponding control hypothesis can be
formulated assuming that the angle between the gun barrel and the vector pointing
from the gun to the target is a controlled variable. Alternative hypotheses can also
be formulated. For example, it is possible that subjects controlled the position of the
pistol in space or the trajectory of the center of mass of the arm with the gun (cf.
Suzuki et al., 1997). Other uncontrolled manifolds can be computed for these alternative control hypotheses.

Subjects were required to perform the task with different initial positions of the
pistol and even in the presence of an elastic restraint of the elbow. The angle between
the gun barrel and the vector pointing from the gun to the target was shown to be an
essential control variable for successful performance. It was controlled from early
in the gun's movement, which started with the gun pointing in a different direction,
not only as the target was approached. This is a rather unexpected result since
successful shooting does not depend on success in stabilizing this angle when the
pistol is pointing away from the target. Although this variable was expected to show
low variability during successful shooting, control of this variable was achieved by
a specific strategy, namely, stabilizing a corresponding UCM.

Similar analyses were performed for two alternative control hypotheses related to a stabilization of the spatial position of the gun and of the center of mass of the arm with the gun. For these two hypotheses, the differences between the two components of the variance (parallel and perpendicular to corresponding manifolds) were small or absent during certain movement phases. For example, the center-of-mass trajectory was controlled during the first half of the movement time, when large-amplitude motions occurred in the proximal arm joints, but not during the aiming phase immediately prior to shooting.

Taken together, this early experience with using the UCM-hypothesis is encouraging. It shows that the principle of minimal interaction is not an abstraction but can be tested using quantitative methods. Moreover, the UCM-hypothesis suggests a method of testing control hypotheses in cases where a motor task may be associated with stabilization of less obvious variables than in the examples presented.

In particular, more recently, the UCM-approach has been applied to studies of finger coordination in experiments with force production similar to those reviewed earlier (Latash et al., 2001; Scholz et al., 2002). Analysis of components of finger force variance has revealed certain counterintuitive features of finger coordination: When subjects were asked to produce an oscillating pattern of total force at a particular rhythm, patterns of individual finger force changes across cycles were more likely to show force destabilization rather than stabilization. In other words, if one finger showed a larger-than-average force at a particular phase in a cycle, another finger was also more likely to show a larger-than-average force leading to an amplification of the original error in the total force. Such patterns were typically compatible with stabilization of the total moment produced by the fingers with respect to a midpoint between the two lateral fingers.

Most everyday tasks involving coordinated action by the fingers impose only mild constraints on total force. For example, when a person grips a spoon, the grip force should be above the threshold for slipping. In contrast, the moment needs to be controlled precisely because an error would lead to spoon rotation and spilling of its contents. It is possible that multi-finger synergies developed during the lifetime are biased toward moment stabilization, and this bias is reflected in our findings.

Possible Implications for Other Areas of Biology

The main ideas of structural unit organization are applicable to different biological objects ranging from macromolecules to societies. In particular, the principle of abundance, together with axioms 1 and 2 of structural unit organization, has a very direct relation to an impressive series of studies of the phenomenon of evolvability by Kirschner and Gerhardt.

In their influential recent series of papers (reviewed in Gerhardt and Kirschner, 1997; Kirschner and Gerhardt, 1998), Kirschner and Gerhardt suggested that organisms possess a feature of evolvability (i.e., a capacity to generate heritable phenotypic variations). Among the important contributors to evolvability mentioned by these authors are weak linkages among protein elements, redundancy, and exploratory behavior. Note that modifiable ("weak") linkages among elements and redundancy (we prefer to call it abundance) are important features of structural units. Their presence allows the realization of many states or configurations of elements, providing a basis for the selection of particular functional outcomes based on additional criteria. Using this structural unit organization of the intracellular machinery, an organism is able to reduce the normally present constraints on change and facilitate mutations of somatic cells, leading to the accumulation of nonlethal variation.

Cells have a system of communication with the external world (a cell language) that includes signals, receptors of the signals, pathways transmitting the signals inside the cell to target structures, and reactions of the target structures. Almost all signals are molecules, not waves. These signals are conditional; that is, they induce reactions, not by direct chemical reactivity of their molecules but by their noncovalent weak binding to receptor molecules. Each cell has a characteristic set of receptors to many types of signal molecules, which define its vocabulary.

As a result, target reactions are induced by an interaction among many elements ("words") of the cell language that may be viewed as a structural unit (Vasiliev et al., 1969). These reactions are usually complex "synergies" such as proliferation (including activation of DNA synthesis, mitosis, etc.) or motility (including activation of pseudopodial extension, attachment of these pseudopodia, contraction, etc.). Each target synergy (e.g., activation of proliferation) may be induced by various combinations/sequences of signals (cf. principle of abundance). Possibly, one of the integrating mechanisms acts via cytoskeleton, particularly via tensions in various parts of the cells developed by the actin-myosin cortex and modified by the microtubular system (Glushankova et al., 1997; Krendel et al., 1999). This organization is another example of the principle of nonindividualized control.

A protein molecule may be viewed as the smallest living being built of structural units. Each structural unit (a "word" in terms suggested by Gelfand and Kister [unpublished]) represents a three-dimensional part of an amino acid sequence. There are three main types of structural units: helices, strands, and loops. Lawful sequences of "words" form many possible languages of proteins. Gelfand and Kister have studied one such language, namely, the language of immunoglobulin antibodies. "Phrases" within this language usually contain 21 words. Thousands of possible phrases have an important feature: Words at the same position within the phrases (e.g., all the first words or all the eleventh words) consist of a relatively small number of amino acid patterns ("keywords"). Moreover, keywords of different words within a phrase are interdependent so that, for example, the keyword makeup of the first word makes certain keyword makeups of the second word more probable. These "rules of grammar" limit the hypothetically possible word sequences without strictly defining them. This organization resembles the organization of natural languages and presents an illustration of the principle of nonindividualized control. We believe that deciphering the language of proteins is a step toward formulation of an adequate language for biological objects.

Conclusion

The studies reviewed in this chapter show that the idea of structural units, combined with the principle of minimal interaction, creates a fruitful framework for an analysis of the cooperation among numerous elements contributing to movement production. We suggest that the formulation of the problem of motor redundancy should be reconsidered. It is more adequate to view structural units underlying everyday movements as abundant (i.e. rich in elements that allow movements to be adaptive, flexible, and stable against both external perturbations and intrinsic, spontaneously occurring errors). This principle of abundance is reflected in a particular approach to the problem of motor coordination, the uncontrolled manifold hypothesis. This hypothesis and the associated toolbox have been successfully applied to a variety of motor tasks including whole-body movements, multijoint limb movements, and multi-finger force production. Examples from other areas of biology considered in the last section suggest that the notion of structural units and the principle of abundance may be applicable to a variety of problems of biological coordination.

Acknowledgments

We are deeply indebted to A. Kister, V. Retakh, and Y. Vasiliev for many stimulating discussions. Comments by Dagmar Sternad on an earlier version of this chapter are deeply appreciated.

References

Abbs, J.H., and Gracco, V.L. (1984) Control of complex motor gestures: Orofacial muscle responses to load perturbations of the lip during speech. *J Neurophysiol* 51: 705-723.
Atkeson, C.G. (1989) Learning arm kinematics and dynamics. *Ann Rev Neurosci* 12: 157-183.
Berkinblit, M.B., Feldman, A.G., and Fookson, O.I. (1986a) Adaptability of innate motor patterns and motor control mechanisms. *Behav Brain Sci* 9: 585-638.
Berkinblit, M.B., Gelfand, I.M., and Feldman, A.G. (1986b) A model for the control of multijoint movements. *Biofizika* 31: 128-138.
Bernstein, N.A. (1923) Studies on the biomechanics of hitting using optical recording. *Annals of the Central Institute of Labor,* v. 1, pp. 19-79 (in Russian).
Bernstein, N.A. (1927) Kymozyclographion, ein neuer Apparat fur Bewegungsstudium. *Pflugers Arch ges Physiol Menschen und Tiere* 217: 783-793.
Bernstein, N.A. (1935) The problem of interrelation between coordination and localization. *Arch Biol Sci* 38: 1-35 (in Russian).
Bernstein, N.A. (1947) *On the construction of movements.* Moscow: Medgiz (in Russian).
Bernstein, N.A. (1967) *The co-ordination and regulation of movements.* Oxford: Pergamon Press.
Bernstein, N.A., and Popova, T.S. (1929) Untersuchung uber die Biodynamik des Klavieranschlagers. *Arbeitsphysiol Ztschr Physiol Menschen Arbeit und Sport* 1: 396-432.
Cordo, P.J., and Nashner, L.M. (1982) Properties of postural adjustments associated with rapid arm movements. *J Neurophysiol* 47: 287-302.
Danion, F., Latash, M.L., Li, Z.M., and Zatsiorsky, V.M. (2000) The effect of fatigue on multi-finger coordination in force production tasks. *J Physiol* 523: 523-532.

Danion, F., Latash, M.L., Li, Z.M., and Zatsiorsky, V.M. (2001) The effect of a fatiguing exercise by the index finger on single- and multi-finger force production tasks. *Exp Brain Res* 138: 322-329.

Dietz, V., Quintern, J., and Berger, W. (1984) Corrective reactions to stumbling in man: Functional significance of spinal and transcortical reflexes. *Neurosci Lett* 44: 131-135.

Dufosse, M., Hugon, M., and Massion, J. (1985) Postural forearm changes induced by predictable in time or voluntary triggered unloading in man. *Exp Brain Res* 60: 330-334.

Feldman, A.G., and Levin, M.F. (1995) Positional frames of reference in motor control: Their origin and use. *Behav Brain Sci* 18: 723-806.

Fookson, O.I., Berkinblit, M.B., and Feldman, A.G. (1980) The spinal frog takes into account the scheme of its body during the wiping reflex. *Science* 209: 1261-1263.

Forssberg, H. (1979) Stumbling corrective reaction: A phase dependent compensatory reaction during locomotion. *J Neurophysiol* 42: 936-953.

Gelfand, I.M. (1991) Two archetypes in the psychology of man. *Nonlinear Sci Today* 1: 11-16.

Gelfand, I.M., and Latash, M.L. (1998) On the problem of adequate language in movement science. *Motor Control* 2: 306-313.

Gelfand, I.M., and Tsetlin, M.L. (1962) On certain methods of control of complex systems. *Adv Math Sci* 17: 103 (in Russian).

Gelfand, I.M., and Tsetlin, M.L. (1966) On mathematical modeling of the mechanisms of the central nervous system. In Gelfand, I.M., Gurfinkel, V.S., Fomin, S.V., and Tsetlin, M.L. (Eds.), *Models of the structural-functional organization of certain biological systems,* pp. 9-26, Moscow: Nauka (in Russian; translation available in 1971 edition by MIT Press, Cambridge, MA).

Gerhardt, J., and Kirschner, M. (1997) *Cells, embryos and evolution.* London: Blackwell Scientific.

Gielen, C.C.A.M., Ramaekers, L., and van Zuylen, E.J. (1988) Long-latency stretch reflexes as co-ordinated functional responses in man. *J Physiol* 407: 275-292.

Glushankova, N.A., Alieva, N.O., Krendel, M.F., Bonder, E.M., Feder, H.H., Vasiliev, J.M., and Gelfand, I.M. (1997) Cell-cell contact changes the dynamics of lamellar activity in non-transformed epitheliocytes but not in their ras-transformed descendants. *Proc Natl Acad Sci USA* 94: 879-883.

Grillner, S. (1979) Interaction between central and peripheral mechanisms in the control of locomotion. In Granit, R., and Pompeiano, O. (Eds.), *Reflex control of posture and movement,* pp. 227-235. Amsterdam, New York, Oxford: Elsevier.

Johansson, R.S., and Westling, G. (1984) Roles of glabrous skin receptors and sensorimotor memory in automatic control of precision grip when lifting rougher or more slippery objects. *Exp Brain Res* 56: 550-564.

Kelso, J.A.S. (1995) *Dynamic patterns: The self-organization of brain and behavior.* Cambridge: MIT Press.

Kirschner, M., and Gerhardt, J. (1998) Evolvability. *Proc Natl Acad Sci USA* 95: 8420-8427.

Krendel, M.F., Glushankova, N.A., Bonder, E.M., Feder, H.H., Vasiliev, J.M., and Gelfand, I.M. (1999) Dynamics of contacts between lamellae of fibroblasts: Essential role of actin cytoskeleton. *Proc Natl Acad Sci USA* 96: 9666-9670.

Latash, M.L. (1993) *Control of human movement.* Champaign, IL: Human Kinetics.

Latash, M.L. (1996) How does our brain make its choices? In Latash, M.L., and Turvey, M.T. (Eds.), *Dexterity and its development,* pp. 277-304. Mahwah, NJ: Erlbaum.

Latash, M.L. (2000) There is no motor redundancy in human movements. There is motor abundance. *Motor Control* 4: 257-259.

Latash, M.L., Gelfand, I.M., Li, Z.M., and Zatsiorsky, V.M. (1998a) Changes in the force sharing pattern induced by modifications of visual feedback during force production by a set of fingers. *Exp Brain Res* 123: 255-262.

Latash, M.L., Li, Z.M., and Zatsiorsky, V.M. (1998b) A principle of error compensation studied within a task of force production by a redundant set of fingers. *Exp Brain Res* 122: 131-138.

Latash, M.L., Scholz, J.F., Danion, F., Schöner, G. (2001) Structure of motor variability in marginally redundant multi-finger force production tasks. *Exp Brain Res* 141: 153-165.

Li, Z.M., Latash, M.L., Newell, K.M., and Zatsiorsky, V.M. (1998b) Motor redundancy during maximal voluntary contraction in four-finger tasks. *Exp Brain Res* 122: 71-78.

Li, Z.M., Latash, M.L., and Zatsiorsky, V.M. (1998a) Force sharing among fingers as a model of the redundancy problem. *Exp Brain Res* 119: 276-286.

Li, Z.M., Zatsiorsky, V.M., and Latash, M.L. (2000) Contribution of the extrinsic and intrinsic hand muscles to the moments in finger joints. *Clin Biomech* 15: 203-211.

Mussa-Ivaldi, F.A., Morasso, P., and Zaccaria, R. (1989) Kinematic networks. A distributed model for representing and regularizing motor redundancy. *Biol Cybern* 60: 1-16.

Nashner, L.M. (1976) Adapting reflexes controlling human posture. *Exp Brain Res* 26: 59-72.

Nashner, L.M., and Woollacott, M. (1979) The organization of rapid postural adjustments of standing humans: An experimental-conceptual model. In Talbott, R.E., and Humphrey, D.R. (Eds.), *Posture and movement,* pp. 243-257. New York: Raven Press.

Newell, K.M., and Corcos, D.M. (1993) *Variability in motor control.* Champaign, IL: Human Kinetics.

Prilutsky, B.I. (2000) Coordination of one- and two-joint muscles: Functional consequences and implications for motor control. *Motor Control* 4: 1-44.

Rosenbaum, D.A., Loukopoulos, L.D., Meulenbroek, R.G.M., Vaughan, J., and Engelbrecht, S.E. (1995) Planning reaches by evaluating stored postures. *Psychol Rev* 102: 28-67.

Saltzman, E.L., and Munhall, K.G. (1989) A dynamical approach to gestural patterning in speech production. *Ecol Psychol* 1: 333-382.

Scholz, J.P., Danion, F., Latash, M.L., Schöner, G. (2002) Understanding finger coordination through analysis of the structure of force variability. *Biol Cyber* 86: 29-39.

Scholz, J.P., and Latash, M.L. (1998) A study of a bimanual synergy associated with holding an object. *Hum Move Sci* 17: 753-779.

Scholz, J.P., and Schöner, G. (1999) The uncontrolled manifold concept: Identifying control variables for a functional task. *Exp Brain Res* 126: 289-306.

Scholz, J.P., Schöner, G., and Latash, M.L. (2000) Identifying the control structure of multijoint coordination during pistol shooting. *Exp Brain Res* 135: 382-404.

Schöner, G. (1995) Recent developments and problems in human movement science and their conceptual implications. *Ecol Psychol* 8: 291-314.

Seif-Naraghi, A.H., and Winters, J.M. (1990) Optimized strategies for scaling goal-directed dynamic limb movements. In Winters, J.M., and Woo, S.L.-Y. (Eds.), *Multiple muscle systems. Biomechanics and movement organization,* pp. 312-334. New York: Springer-Verlag.

Stein, R.B. (1982) What muscle variable(s) does the nervous system control in limb movements? *Behav Brain Sci* 5: 535-577.

Suzuki, M., Yamazaki, Y., Mizuno, N., and Matsunami, K. (1997) Trajectory formation of the center-of-mass of the arm during reaching movements. *Neurosci* 76: 597-610.

Vasiliev, J.M., Gelfand, I.M., Guberman, S.I., and Shik, M. (1969) Interactions in biological systems. *Priroda* [Nature] 6: 13-21 and 7: 24-33 (in Russian).

Zatsiorsky, V.M., Li, Z.M., and Latash, M.L. (1998) Coordinated force production in multi-finger tasks: Finger interaction and neural network modeling. *Biol Cybern* 79: 139-150.

Zatsiorsky, V.M., Li, Z.M., and Latash, M.L. (2000) Enslaving effects in multi-finger force production. *Exp Brain Res* 131: 187-195.

12
CHAPTER

Bernstein Versus Pavlovianism

An Interpretation

Onno G. Meijer
Faculty of Human Movement Sciences,
Vrije Universiteit

From July 31 through August 7, 1948, the Lenin Academy of Agricultural Sciences met in Moscow (Regelmann, 1980; cf. Lysenko, 1954). According to some, it was a "carefully prepared" (Bongaardt and Meijer, 2000, p. 59) coup to bring Lysenkoism to power and get rid of Mendelism in genetics. Others describe the meeting as an honest but failed attempt to attack Lysenko for his unfounded inheritance of acquired characters and to bring Soviet genetics back to the international scene (Graham, 1987). For Lysenko's adversaries, the meeting turned into a disaster, with respectable geneticists denouncing their own perfectly scientific views (cf., Regelmann, 1980). Lysenko had won, and Mendelism (or standard genetics) became forbidden so that Soviet and somewhat later Chinese genetics became irrelevant to the rest of the world for more than a decade. Western intellectuals, particularly in France (cf. Judson, 1979), turned against communism, or at least Stalin's version of it.

The Pavlovian Affair

From June 28 through July 4, 1950, a similar meeting took place, this time sponsored by the Academy of Sciences of the U.S.S.R. and the Academy of Medical Sciences

(Akademija Nauk S.S.S.R. and Akademija Meditsinskikh Nauk S.S.S.R., 1950). Sergei Vavilov—brother of Nikolai, the geneticist, whom the system had destroyed in 1940—sounded the alarm to protect the integrity of Pavlov's heritage:

> *There have been attempts—not too frequent, happily—at an erroneous and unwarranted revision of Pavlov's views. . . . Strange and surprising though it may seem, the broad Pavlov road has become little frequented, comparatively few have followed it consistently and systematically. Not all our physiologists have been able, or have always been able, to measure up to Pavlov's straight-forward materialism. . . . The time has come to sound the alarm. (Sergei Vavilov, 1950, quoted from Graham, 1987, p. 175)*

Attacks at the 1950 meeting were mainly directed at Anokhin, Beritashvilli, and Orbeli (cf. Feigenberg and Meijer, 1999), but Luria was also in trouble and Nikolai Aleksandrovich Bernstein had not even been invited to participate (Pickenhain, 1998a); immediately after the session, Bernstein was fired, to remain haunted by the spectre of Pavlovianism for the rest of his life (Bongaardt and Meijer, 2000).

When Stalin died and Khrushchev came to power (1953), new openings were created, and the lives of scientists generally became more relaxed. Nevertheless, Khrushchev was known to admire Lysenko (Graham, 1987), perhaps not because of the latter's scientific merits but because of the promise, inherent in Lysenkoism, of increasing agricultural output. Although never completely rehabilitated, Bernstein was allowed to work again. Quite naturally, then, there was a feeling of excitement when another meeting took place, the All-Union Conference on Philosophic Questions of Higher Nervous Activity and Psychology, convened by the Academy of Sciences of the U.S.S.R., the Russian Academy of Pedagogical Sciences, and the Ministries of Higher Education of both the U.S.S.R. and the Russian Republic (Graham, 1987).

During the conference, May 8 through 11, 1962, sharp differences of opinion emerged. Although Bernstein was now recognized as the most outspoken critic of Pavlov's theory of reflexes, he was allowed to present his own ideas (cf. Feigenberg and Meijer, 1999). His main point was that in Pavlov's reflex theory, the organism is passive and can only reach equilibrium with its environment, whereas in reality, it acts on it. The Pavlovians disagreed with almost everything he said but reacted with restraint and chose unknown Lekhtman, from the Leningrad Institute for Physical Culture, to attack Bernstein's views:

> *In our opinion, there are serious methodological problems with the conception of [Bernstein's] 'physiology of activity'. This tendentious confrontation between the 'physiology of activity' and the physiology of reflexes, the theory of higher nervous activity, reveals, against the will of its author, the undeniable superiority of the latter, both regarding to facts and to methodology. (cited from Lekhtman, 1963, p. 599, translated by Ines M. Rubin)*

Bernstein was not impressed, and according to Graham (1987), the general feeling of the conference was that of a truce, or even a merger, rather than a continuing conflict. Nevertheless, although never again as sharp as in 1950, the controversy would actually go on until the present day.

In his 1965 "On the Roads Towards a Biology of Activity," Bernstein (1988a) sharply attacked the "epigones" of Pavlov for their distortion of Pavlov's original views:

> The epigones of the theory of I.P. Pavlov distorted the image of this outstanding, world famous scientist in that they turned his theory after his death into dogma. By creating a dogma and insisting on its infallibility—which always, since the times of Aristotle, results in limiting the progress of science—the pupils and followers of I.P. Pavlov have done serious damage to our national science in two different ways. In the first place, the price for their stubborn fixation with long obsolete positions is that practical applications of their views . . . led to insurmountable failures, in psychiatry as well as pedagogics, and even in linguistics. . . . On the other hand, the developments led to all that damage which always occurs in science when the argument switches from trying to convince to using actual violence. This is trivial and deserves no further explanation. (Bernstein, 1988a, p. 242, translated by Lothar Pickenhain)

Clearly, in 1965, the Pavlovian affair was not over.

The Lysenko affair has received relatively wide attention in the international literature (cf., e.g., Darlington, 1949; Huxley, 1949; Joravsky, 1970; Regelmann, 1980). Its counterpart in movement science, however, remained relatively forgotten (for exceptions, see Graham, 1987; Kozulin, 1984; and Mecacci, 1979), apart from the memories of those who were, or are, actually inspired by Bernstein (e.g., Bongaardt and Meijer, 2000; Feigenberg and Latash, 1996; Pickenhain, 1998a; Sirotkina, 1995; and, more implicitly, Gel'fand et al., 1971; Pickenhain and Schnabel, 1988). On the face of it, this historiographical difference between the two affairs appears to be due to the fact that genetics is hard-core international science, whereas the study of movement is still somewhat of an oddity. In general, such lack of recognition tends to lead to mythologizing—that is, to emphasizing the difficulties of bygone days (Powers, 1982).

Time has come, therefore, for an impartial evaluation of the controversy between Bernstein and Pavlovianism. The trouble is, no one is impartial. Personally, I feel incompetent to do the job because I do not master the Russian language. On the other hand, I have been involved in the historiography of Bernstein's work for several years now (cf., e.g., Bongaardt, 1996, and the section "Bernstein's heritage" in the journal *Motor Control*), and many Russian-speaking friends have been kind enough to share their insights with me. Still, I feel uneasy with the work at hand because my view of the Pavlovian affair does differ from that of many good friends I interviewed. But then, they are used to dissidents.

First, I will summarize Pavlov's life and work insofar as it is relevant to the present analysis. I will then briefly present the apparent context of the Pavlovian affair (i.e., dialectical materialism). Subsequently, I will describe Bernstein's attack on Pavlovianism and the Pavlovian attack on Bernstein, and finally entertain that burning question: What was it all about?

Ivan Petrovich Pavlov

Ivan Petrovich Pavlov (1849-1936) was an outstanding scientist. A wealth of literature, written by his friends as well as his enemies, serves to illustrate this indubitable fact (Adrian, 1936; Asratyan, 1953; Babkin, 1974; Barcroft, 1936; Editorial, 1936; Frolov, 1937; Gray, 1979; Grigorian, 1974; Hill, 1936; Pickenhain, 1998b, 1998c; Solandt, 1935; Starling, 1925; Todes, 1995; Windholz, 1983, 1990; for other references, see the Cumulative Bibliographies of the journal *ISIS*).

A country kid from Ryazan, son of a priest, he opted for natural sciences in 1870, becoming part of that vast materialist movement in Russia that wanted to improve society through science (cf. e.g., Kosmodemyansky, 1956; Pickenhain, 1998a, 1998c). In 1904, Pavlov received the Nobel Prize in medicine for his work on digestion: From the time that food is observed, through its final passage in the intestines, specific digestive glands are stimulated. We owe this basic textbook fact to Pavlov, who himself used it to make an abrupt switch to anticipation in the nervous system:

> *Of course, [during my experiments] I could not fail to notice that stimulation of the digestive glands which at the time was regarded as 'psychical', i.e., the fact that in hungry animals or in man, the seeing of food, talking of food, or even thinking of food leads to the production of saliva. (Pavlov, 1923, quoted from Pickenhain, 1998d, p. 13, translated from the German by this author)*

Pavlov found that when a dog is exposed repeatedly to the sound of a bell before it is offered food, the sound of the bell is sufficient to stimulate the production of saliva: a *conditioned reflex*. He would be engaged in the study of conditioned reflexes from 1902 to the end of his life.

During the Russian revolution, in 1917, Pavlov was a world-famous man, already 68 years old, working on a materialist theory of the mind. During the years of the Red Terror, 1918-1920, he was horrified about "the death of our homeland" (Todes, 1995, p. 383), lost a son who wanted to join the White army, while another son who had already joined the White Resistance was to disappear for a long time. Clearly, the Pavlov family were enemies of the state, and Pavlov announced that he wanted to emigrate. Thus arose that "combative collaboration" (Todes, 1995) between Pavlov and the Bolsheviks that would be so beneficial to both.

In 1921, it had become clear that Bolshevism would win, and Lenin occupied himself with the question of how to establish the new state. Realizing that he needed science and international prestige, he decided to produce an extraordinary decree: "On Conditions Facilitating the Scientific Work of Academician I.P. Pavlov and his Coworkers," assuring Pavlov a privileged place in Soviet science (Todes, 1995). Under the new economic policy, the system relaxed for a few years, but after Lenin's death (1924), a power struggle emerged and Stalin came to power. The first signs that things were changing for the worse came in 1928-1929, when the Soviet authorities started to interfere with membership of the Academy of Sciences. Pavlov was furious and became a symbol of resistance. In December 1929, he used the celebration of the 100th anniversary of Sechenov's birth to express his feelings. Addressing Sechenov's portrait, he exclaimed:

> *Oh noble and stern apparition! How you would have suffered if in living human form you still remained among us! We live under the rule of the cruel principle that the state and authority is everything, that the person, the citizen is nothing. Life, freedom, dignity, convictions, beliefs, habits, the possibility of studying, means for life, food, housing, clothing—everything is in the hands of the government. . . . Naturally, gentlemen, the entire citizenry is transformed into a quivering, slavish mass. . . . On such a basis, gentlemen, not only can no civilized state be built, but no state at all can long survive. (Pavlov, 1929, quoted from Todes, 1995, p. 400)*

The audience was shocked, but 80-year-old Pavlov was about to show a remarkable change of mind.

It may have been his friendship with some dedicated communists or his realization that his laboratory completely depended on the government, but in all probability it was his fear of Hitler that made him switch to praising the Soviet government. In 1934, while the coming terror of Stalin's purges was not yet fully visible, he clearly sided with the Soviets against Hitler: "We should especially sympathize with and facilitate our government's struggle for peace" (Pavlov, 1934, quoted from Todes, 1995, p. 412). In 1935, Pavlov was president of the 15th International Congress of Physiologists, held in Russia. He ended his opening words with the following:

> *In conclusion, we, Russian physiologists, wish to express gratitude to our government which has enabled us to receive our esteemed guests in a worthy manner.* (Applause) *(Pavlov, 1955e, p. 58)*

So, when Pavlov died, in 1936, the scene was set for "a generation of fully Soviet heirs" (Todes, 1995, p. 418). Or was it?

Scientific Controversies

Pavlov had strong feelings of personal integrity, and his coworkers trusted him and liked him very much, almost without exception (Windholz, 1990). Still, being driven around in a limousine or having a street displaced because the noise bothered his dogs (Todes, 1995) may have failed to turn him into a source of moral inspiration for his colleagues in the Soviet Union. Moreover, he had a liking for spicy, belligerent commentary and was involved in several scientific controversies. I will limit myself to his disagreement with Sherrington and his dislike of *Gestalt*.

In 1933, a booklet by Sir Charles Sherrington appeared titled *The Brain and Its Mechanism*. In this lecture at the University of Cambridge, Sherrington emphasized that science does not understand the relationship between mind and brain, ending with the rather puzzling remark that maybe it shouldn't. Because, if it did: "We need not be prophets to foresee that then will come the long-told speedy extinction of man" (Sherrington, 1933, p. 35). Pavlov was flabbergasted: "What do you think of that? What does it mean? Why, it's simply preposterous!" (Pavlov, 1955b, p. 564).

Pavlov is particularly puzzled when Sherrington (1933, p. 33) sighs: "If nerve-activity have relation to mind." Pavlov:

"If nerve activity have relation to mind. . . ." I did not trust my knowledge of English and so I requested others to translate it for me.

How can it be that at the present time a physiologist should doubt the relation between nervous activity and the mind? This is the result of a purely dualistic concept. This is the Cartesian viewpoint, according to which the brain is a piano, a passive instrument, while the soul is a musician, extracting from this piano any melodies it likes. . . .

Gentlemen, can anyone of you, who has read Sherrington's booklet, say anything in defence of the author? I believe that this is not a matter of some kind of misunderstanding, thoughtlessness or misjudgement. I simply suppose that he is ill, although he is only seventy years old, that these are distinct symptoms of old age, of senility. . . . Take my wife, for example. She is an obvious dualist. She is religious, but at the same time her attitude to things is not distorted. (Pavlov, 1955b, pp. 563 and 564-565)

Sherrington (1933, 1952, 1955, 1979) rarely mentioned Pavlov in his works, but when he did so, it was with praise. In 1932, he had received the Nobel Prize in medicine. He was to live until 1952, so he cannot have been that ill in 1933. Still, Pavlov rightly unmasks him as a closet dualist, as becomes clear from Sherrington's "Brain Collaborates With Psyche" (1955, p. 217), in which he states: "the cortex is the region where brain and mind meet" (ibid.). And for Pavlov, ardent materialist as he was, dualism (or idealism) was anathema.

Pavlov's language becomes even more juicy when he discusses *Gestalt*. Let there be no doubt: "Perception, if considered profoundly, is simply a conditioned reflex" (Pavlov, 1955a, p. 574). Thus, *gestaltists* have it all wrong, and:

These gentlemen should have made a proper study of physiology, i.e., they should have thoroughly read Helmholtz. But instead, they content themselves with a play on words. . . . Apparently, Gestaltists are recruited from among very superficial persons. Such, for instance, is Professor Kurt Lewin of Berlin university—don't trifle with this! Lewin . . . affirms that . . . stimuli do not produce any action. . . . He illustrates this point of view by splendid experiments—observations carried out on himself. All the experiments and observations of the teachers and of the pupils are cited in the book. So you can imagine what an intellectual beauty it is! (ibid., pp. 574-575)

Why would Pavlov be so adamant against *Gestalt?* Apparently, because it hit him in the heart. Even by his friends, the American behaviorists (cf. Windholz, 1983), he had been criticized for paying too little attention to the organization of the brain. Pavlov had been shocked, emphasizing that the "theory of conditioned reflexes undoubtedly established in physiology the fact of *temporary* connection between diverse . . . stimuli with certain units of activity of the organism" (Pavlov 1955d, pp. 427-428, my italics) and that he had introduced the notion of a *mosaic* of cortical functions (1955a, p. 571), responsible for the *dynamic stereotypy* of complicated actions (1955c, p. 448; cf. Asratyan, 1953, p. 118). Thus, in the 1930s, it looked as if Pavlov had really accepted a dynamical view of the organization of the brain:

Of course, man is a system (more bluntly, a machine) that is subjected, as any other natural system, to the inexorable, unified laws of all nature. But within the horizon of our contemporary scientific understanding, it is a system that is unique because of its capability of the highest degree of self-regulation . . . preserving itself, repairing itself, correcting itself, and even perfecting itself. The most telling, the strongest and remaining impression in the study of higher nervous activity with our methods, arises from the extraordinary plasticity of this activity, and its fantastic possibilities. Nothing remains static, unresponsive, but anything can be reached at all times, anything can change itself for the better, if only the necessary conditions are realized. (Pavlov, 1932, quoted from Pickenhain, 1998d, pp. 156-157; my translation from the German)

However impressive such statements were, Pavlov's last-minute changes failed to help: By the early 1930s, the international scientific community came to the conclusion that Pavlov tried to understand the dynamic organization of the brain through the study of semistatic elements. *Gestalt* had shown, for some time already, that this would not work (cf. also Planck, 1910). And Pavlov mastered all his venom to ridicule the *gestaltists*. The most beautiful example I found is his discussion of Köhler's study of the intelligence of chimpanzees. Time and again, Köhler (cf. 1973) depicts how a chimpanzee tries to solve a particular problem, fails, takes a rest or does something else, returns to the problem, and then all of a sudden turns out to be able to solve it. The temptation to see the animal "reflect" during the pause is very great indeed. Pavlov commented:

When the ape is given the task of taking hold of fruit suspended at a certain height, and when for the purpose of accomplishing it he needs definite instruments, for example, a stick and some boxes, all his unsuccessful efforts to get the fruit are not, according to Köhler, proof of intelligence. This is simply the method of trial and error. When the ape becomes tired, as a result of his unsuccessful efforts, he gives up and remains for some time in sitting posture. When he has rested he tries again and succeeds in accomplishing his task. According to Köhler, the ape's intelligence is proved by the fact that he sits for a period without doing anything. He literally says that, gentlemen. In his view the ape accomplishes some kind of intellectual work when it is sitting, and this

proves its intelligence. How do you like it? It turns out that nothing but the
silent inaction of the ape proves its intelligence! (Pavlov, 1955f, pp. 558-559)

For Pavlov, *Gestalt* really was the enemy.

The Pavlov That Was

To me, the most amazing thing about science in the Soviet Union is that it could
flourish amidst the terror.

In order to keep their sanity, many found peace in the normal things of everyday
life (Aksyonov, 1993-1994). After all, one could do one's work in relative isolation
(Orlovsky, personal communication, 1999). And from time to time, one needed a
good story, something of untainted beauty, to be able to keep hope.

Some mythologies of Pavlov depict him as the eternal Bolshevik, who late in his
life came to understand his own true position; others see him as the first consistent
dissident (cf. Todes, 1995). But the real Pavlov was a man of flesh and blood. A man
of world fame, loved by his coworkers, with a strong personal feeling of morale,
prone to compromise with the system in order to keep his laboratory going, later
deciding that the Soviet system was less of a danger than Hitler's national socialism,
and always defending his own scientific views with gusto. I have met with many
admirable people who wanted to keep Pavlov's memory "clean." I think that
the Pavlov that really was is as clean as it comes. So, the summary of his life and
his scientific controversies should not lessen one's esteem for Pavlov. It should
enliven it.

The Apparent Context: Dialectical Materialism

After Pavlov's death, an obituary appeared in the *Izvestiia,* written by Nikolai
Bukharin. Communist Bukharin had been Pavlov's friend for a long time and might
have been instrumental in Pavlov's praising the government in the final years of his
life (Todes, 1995). Bukharin wrote: "Pavlov is entirely ours, and we will never
surrender him to anybody" (Bukharin, 1936, quoted from Todes, 1995). Thus, the
stage was set for the Pavlov affair.

Soviet authorities must have realized that there were several difficulties with
Bukharin's sentiment. First, although Pavlov had been more positively inclined
toward the Soviet government in the last years of his life, he had never been a party
member. Consequently, "Pavlov is ours" was somewhat ambiguous: Was it to mean
"Soviet," which he had never been, or "Russian"? It appears that the latter meaning
is more adequate, but turning internationalist socialism into Russian nationalism
was just on its way, and it would take until the end of World War II (the Great Patriotic
War) before antiforeign sentiments could be used to their fullest. So, Bukharin may
have been thinking "Russian," but it certainly was communist Russia that tried to
own Pavlov's heritage. And therein lies the second, more important problem: the

relationship between Pavlov's theories and dialectical materialism, the obligatory philosophy of science in the Soviet Union since about 1930 (Kozulin, 1984).

Of course, there was no problem with "materialism" since Pavlov had been one of the most outspoken materialists of his era, and his ardent antidualism/antiidealism would cause several leading Western philosophers to reject Pavlov's views altogether (e.g., Popper and Eccles, 1977). No, the problem for the Soviet Union was with "dialectical." How dialectical had Pavlov been? The notion itself derives from 19th-century debates, particularly what Marx and Engels made of Hegel's philosophy. "Dialectical" has to do with motion, development, change, process. In the philosophy of science that became a central characteristic of Stalin's Soviet Union, three laws were supposed to capture "dialectical."

First (cf. Graham, 1987), there was the Law of the Transition of Quantity into Quality. To 20th-century science at large, there was nothing special about this law. It had, in different phrasing, been formulated in thermodynamics (e.g., Planck, 1910), it agreed with the central tenet of *Gestalt,* and it was fundamental in systems theory (e.g., Weiss, 1959). Russian mathematics played a decisive role in its further development (e.g., Andronov and Chaikin, 1949), and to date, it is an important principle in applications of the theory of dynamical systems (e.g., Haken, 1977). As stated earlier, however, Pavlov was never very successful with "process." His "dynamic stereotypy" (1955c) was not something qualitatively new that emerged from the reflexes but needed "profound traces" of stimuli with "precise and constant effects" (ibid.). It was, in other words, the building up of a track record.

The second law (cf. Graham, 1987), that of the Unity and Struggle of Opposites, may have been somewhat easier on Pavlov. After all, for him the brain is a constantly changing balance of excitation (freedom) and inhibition (restraint). Still, in Pavlov's work, it is a rather dull adding and subtracting that the nervous system appears to be doing all the time, with nothing of the excitement about the fact that the presence of opposing forces may lead to dynamically stable states, as in the political theory of Marxism or in the scientifically well-known example of pendulum movement (Bernstein and Popova, 1929; for modern applications in the theory of dynamical systems, see, e.g., Prigogine and Stengers, 1984). As an argument for Pavlov's heirs to show that the master had been a dialectical materialist all the time, the second law could not be used without some unease.

Third (cf. Graham, 1987), the Law of the Negation of the Negation stresses that synthesis results from the struggle between opposites. It is the law that emphasizes historicity (i.e., the fact that things really disappear without a trace). For Pavlov, it was clear that conditioned reflexes can disappear, that the animal can become deconditioned, but the extinction of conditioned reflexes never was an extinction "without a trace." Again, the verdict has to be negative.

In the preceding section on Pavlov's life and work, I presented two scientific controversies. One was with Sherrington and had to do with dualism. For Soviet scientists, Pavlov's antidualism would remain unproblematic. (It would even give the Pavlovians fuel to accuse adversaries of Pavlov.) The second controversy was about organization. I emphasized Pavlov's problems with *Gestalt,* but behaviorists as well, or even Sherrington, were emphasizing that the brain is a dynamically changing organization. Because of his focus on the details, Pavlov's brain theory was never very impressive (Gray, 1979). In fact, Pavlov's brain theory could not be made to agree with the "dialectical" in dialectical materialism. Thus, the heritage that Bukharin so proudly announced was to be a very uncomfortable one to Soviet science, to say the very least. It fell to a dedicated dialectical materialist to attack Pavlov in this respect. Later, this man would be accused of a lack of patriotism.

A Case of Bad Timing: Bernstein Versus Pavlov

Nikolai Aleksandrovich Bernstein (1896-1966) needs no introduction (Bongaardt, 1996; Bongaardt and Meijer, 2000; Feigenberg and Latash, 1996; Gel'fand et al., 1971; Gurfinkel, 1988; Gurfinkel and Cordo, 1998; Luria, 1987; Pickenhain and Schnabel, 1988; Sirotkina, 1995; the section "Bernstein's Heritage" in *Motor Control,* from 1998 onward).

When Bernstein began to work in the Central Labor Institute in Moscow, he enjoyed a lot of freedom (Bongaardt and Meijer, 2000). He worked in biomechanics, made technical improvements in the filming of movements, cooperated with the Institute for Music Sciences, and regularly made his appearance in Kornilov's psychological lab, where he could meet and work together with inspiring colleagues such as Luria and Vygotsky (cf. Feigenberg's 1988 bibliography). Bernstein made use of the then dominant mechanicism (Braune and Fischer, 1895-1904; cf. Meijer and Wagenaar, 1998) and stressed "the high degree of automation . . . mechanical simplicity and lawful structure" of repeated movements (Bernstein, 1927a, p. 789; cf. Bongaardt and Meijer, 2000). In the period 1928-1930, however, when the government started to interfere with membership of the Academy of Sciences and dialectical materialism was decreed to be the starting point for all neuropsychology, Bernstein's theories began to change.

The clearest example of such change can be found in a paper with Ms. T.S. Popova on the biodynamics of piano playing (Bernstein and Popova, 1929; cf. Feigenberg, 1988). The term *biomechanics* is now replaced by *biodynamics,* as if the work is more "dialectical." In a way, it is. Although Bernstein always had a keen eye for mathematics, and even published a mathematical paper on his own (Bernstein, 1927b), Ms. Popova was a real mathematician (Feigenberg, personal communication, 1999), and Bernstein learned from her about new mathematical developments. In their joint paper, the authors express themselves in terms that are close to the theory of dynamical systems (Andronov and Chaikin, 1949):

> *During slow tempos, the movement [of the pianist] consists of isolated impulses; during medium tempos, the movement corresponds to the oscillations of a compound pendulum; during the fastest tempos, there is a transition to forced elastic oscillations of a simple pendulum. (Bernstein and Popova,*

1929, p. 432, translation by Rob Bongaardt; cf. Bongaardt and Meijer, 2000, p. 60)

The 1929 paper with Popova appears to be a prelude to the Bernstein who would revolutionize movement science.

In the same year, 1929, Bernstein published an amazing statement in the *Grand Medical Encyclopedia* about movements: "There are no situations in which muscle shortening is the cause of a movement" (Fel'dman and Meijer, 1999, p. 119), which is conceptually related to Kurt Lewin's rejection of stimuli producing action, so despised by Pavlov (see earlier discussion). A year later, in a paper on the "Coordination of Movements" in the same *Encyclopedia,* Bernstein refers to the relative autonomy of spinal frogs (cf. Pflüger, 1853) and emphasizes that "no movement can be entirely planned from its very beginning" (Beek and Meijer, 1999, p. 5), again as if to announce his later theories.

For about a decade, Bernstein's papers contain an odd mix of his old mechanicistic views and his attempts to reformulate the central questions of movement science. This culminated, for the time being, in his now-famous paper on "The Problem of the Interrelation of Co-ordination and Localization" (Bernstein, 1967a). Even now, however, the paper makes awkward reading. Bernstein first shows mathematically that motor control depends on both the central signal and the actual situation. He then stresses that the time structure of some movements is predetermined. In 1935, he apparently needed predetermination as a stepping-stone to show that the central signal is different every time in repeated movements, depending on their immediate past and other aspects of their context.

In *Gestalt*-inspired terms, Bernstein (1967a) argues that movements are organized as wholes, which implies a *non-univocal* relationship between central signal and actual movement. It is this latter conclusion that leads to the rejection of Pavlov's brain theory: "The conditioned reflex is . . . every time realized by other cells" (Bernstein, 1988b, p. 80, my translation from the German). Invoking Köhler's principle of "equal simplicity," Bernstein claims that organized movement can only depend on the *organization* of the brain and not in any one-to-one manner on the fixed activity of any of its elements. This organization, the taking together of degrees of freedom, *is* coordination—not the effect of motor control (as Sherrington [1952] would have it) but preceding it—that is, making controllability possible (Bernstein, 1996).

It is a brilliant coup, hitting Pavlov in the heart by exposing Pavlov's inability to deal with the dynamic organization of the brain. Nevertheless, Bernstein does not even mention Pavlov in his 1935 paper. Nor does he discuss Pavlov's "temporary" connections, his notion of "mosaic," or the "dynamic stereotype" (see earlier mention). But to the initiated reader, it was clear that Pavlov's brain theory (not the conditioned reflex) was now, for the first time, rejected from inside the Soviet Union.

Bernstein's timing could not have been worse. In 1935, the International Congress of Physiologists was held in Moscow and Leningrad, with Pavlov as president.

At the time, Pavlov was undisputedly the leading physiologist of the world (Solandt, 1935).

In 1936, Bernstein was to try again. He received the proofs for his book on *Contemporary Research in the Physiology of Nervous Processes* (Bernstein, 1936). In the text, Bernstein shows that the idea of the conditioned reflex stems from the 19th-century German physiologist Meinert and not from Pavlov himself (Feigenberg, personal communication, 2000; cf. Sirotkina, 1995). Bernstein then goes on to show that Pavlov's theory is wrong insofar as it claims a fixed relationship between specific behaviors and specific cells in the cerebral cortex:

> *Let us start to say that there are no direct neuronal connections between the peripheral sensory organs and the hemispheres of the cerebrum. . . . Afferent neurones from the skin immediately stop after entering the spinal cord, and have only synaptic contact with other neurones that go up to the brain. Even these don't go directly to the cortex, but . . . to the subcortical nuclei. . . . From there, a third link leads to the cortex. That this is true, not only anatomically but also functionally, is shown by the fact that all skin reflexes remain intact in decerebrated animals. . . . It is interesting to see that spinal reflexes are under clear influence of local signals: For instance, in the 'scratch reflex' of the decerebrated frog that is chemically irritated at a certain point of its skin, the scratching movement of its leg always hits exactly the right spot. Such a possibility to choose a certain kind of reflex coordination proves that the reflex arc is not closed, it is not fixed for ever; to the contrary, coordinated connections arise every time anew between afferent and efferent neurones—another proof in favor of the synaptic relay in the spinal cord. (Bernstein, 1936, p. 57, translated by Ines M. Rubin)*

Bernstein made extensive corrections to the proofs, but by the time he was ready, Pavlov died, and Bernstein decided to refrain from publishing the book because "now he cannot defend himself" (quoted from the 1999 memory of I.M. Feigenberg, with whom Bernstein discussed these events in the 1960s). So it came to happen that Bernstein's attack on Pavlov aborted.

Bernstein's Motives

If one accepts Kozulin's (1984) suggestion (cf. also Bongaardt and Meijer, 2000) that political events were the major hallmarks in Soviet scientific careers, it becomes important to analyze the events around 1928-1930. In 1928, the government began to bother the Academy of Sciences (which led to Pavlov's extreme anger, as noted earlier). In 1929, "dialecticians" had a first victory over mechanicists, and dialectical materialism became the official philosophy for neuropsychology in the 1930 Congress on Human Behavior (Kozulin, 1984). In the early thirties, Luria had to admit his "mistakes" (Kozulin, 1984, p. 22). So, in the late twenties, it was wise to evaluate one's own philosophical inspiration.

Let us assume that Bernstein did so in 1928 (cf. Bongaardt and Meijer, 2000). What was there to see? Bechterev had died in 1927, possibly killed by the Kremlin (Kozulin, 1984). Although his school was still dominant, its mechanicism (Pickenhain, 1998a) made it an unlikely winner. Indeed, from 1930, Bechterev's theory started to lose influence. Then there was Kornilov, who took Marxism to its extreme in a psychology "where there are no objects, but only processes, where everything is dynamic and timely, where there is nothing that is static" (Kornilov, 1924, quoted from Graham, 1987, p. 164). Kornilov may have appeared to be a safe bet, but he was overdoing it—totally against Bernstein's taste. In 1931, Kornilov was denounced (Kozulin, 1984) and lost his job. Finally, there was Pavlov, world famous, but at the time ardent anticommunist (see earlier discussion). Moreover, Pavlov was unable to deal with "process" and thus a bad dialectician.

For a promising young scientist such as Nikolai Bernstein, knowing that the bells would toll very soon, wouldn't it have been very human for him to dream, in 1928, that he could grab Pavlov's power? Bernstein certainly liked to be recognized, but grabbing power would be too much out of character. Still, around 1928, he met Ms. Popova, who brought him an exciting new view of biodynamics (oscillations, pendulum movements). Afterward, Bernstein would cite Popova's work in a large number of his own papers. She was to marry his brother, and apart from World War II, they lived in the same big house in Moscow (Feigenberg, personal communication, 1999). After Nikolai's death, Popova turned her condominium into a kind of Bernstein museum. Wouldn't it have been very human for Bernstein to see in Ms. Popova's admiring eyes the challenge to get the recognition he was due? No research will be sufficient to establish the if and when of such admiring eyes, but I venture to conclude that the coincidence of the shift in government policy and Bernstein's meeting with Ms. Popova led him on the track that would later turn him into the most formidable adversary of the Pavlovians.

Was Bernstein unfair to Pavlov in his 1935 (and aborted 1936) attack? Slightly so, but without the sharpness of tongue that was so typical of Pavlov himself. In 1935, Bernstein really should have discussed Pavlov's counterarguments to the criticism of his not taking organization seriously, such as the temporary nature of connections, the notion of mosaic, and the dynamic stereotype. Moreover, in 1936, Bernstein accused Pavlov both of plagiarism and of having the wrong theory. Tactically, he should have chosen between the two. Historically, what he did is not exceptional. (Correns accused De Vries of plagiarism and of having it wrong in 1900, thereby creating the Mendelian Revolution; cf. Meijer, 1985.) All in all, Bernstein's publications suggest that he first decided to attack Pavlov, then to look for the best argument to do so.

Recently, another paper was translated that shows traces of Bernstein's failure to get recognition—a 1964 paper on Tsiolkovskii with two blatant mistakes in it, probably "caused" by Bernstein's frustration about his not being officially involved in the training of cosmonauts (Meijer and Feigenberg, 2000). Bernstein rarely made mistakes in his papers, but his 1935 emphasis on the predetermination of the time structure of a movement is clearly mistaken (cf. Bongaardt and Meijer, 2000); it is not consistent with what he said in an earlier paper, or even with an earlier part of the same paper. Bernstein was a man of great personal integrity, and the two times that we know of when he was trying (and failed) to get more recognition reveal his inner unease in the awkwardness of the papers and in the mistakes he made.

The Bernstein That Was

The one thing that appears to be certain is the role of the government in increasing the importance of dialectical materialism in the late 1920s.

The gist of all of the foregoing, therefore, is that Bernstein could attack Pavlov because he was a better dialectical materialist. Some may have hoped that Bernstein was an early dissident. He was not. The real Bernstein was a man of flesh and blood. He created a revolution in movement science, was loved by his coworkers, had a strong personal feeling of integrity, was touched (I think) by the admiration of Ms. Popova, worked tenaciously on his attacks on Pavlovianism, took pleasure in exposing the Pavlovians at a time when the affair had become less urgent, and was a great source of inspiration. I have met with many admirable people who wanted to keep Bernstein's memory "clean." I think that the Bernstein that really was is as clean as it comes. So the summary of his initial role in the Pavlov affair should not lessen one's esteem for Bernstein. It should enliven it.

The Timing of a Bad Case: Pavlovians Versus Bernstein

Bernstein continued to present new arguments against Pavlov's theory of the brain. In line with general developments in the Soviet Union, he dropped *Gestalt*, seen by many as a bourgeois attempt to understand the inner experience of the upper middle class (Ash, 1995). Bernstein himself switched to Weiss's (cf. 1959) hierarchical theory of systems. After the war, he partook in that general atmosphere of optimism that then characterized Moscow (Aksyonov, 1993-1994). Bernstein presented a dynamic view of the brain (Sporns et al., 1998) and came with a more functional hierarchy than Weiss's in his book *On the Construction of Movements* (Bernstein, 1947), which won him the Stalin Prize. Finally, he had done it.

By that time, the Lysenko affair was already brewing, and after the 1948 meeting of the Lenin Academy of Agricultural Sciences, it was clear that dialectical materialism no longer counted. It was pure, unmitigated power that offered itself, to be obtained by pleasing Stalin and his lackeys. Notwithstanding Bernstein's Stalin Prize, or maybe because of it, attacks on his book followed immediately. As Sirotkina describes:

> *Bernstein's monograph was the target of a 'critical review' at an expanded [1948] meeting of the Scientific-Methodological Council of the All-Union Committee of Physical Culture and Sports Affairs, during which the director of the Institute of Physical Culture, I.A. Kriachako, stated that "Professor*

N.A. Bernstein's valuable and original monograph presents a profoundly erroneous characterization of the scientific creativity of the brilliant Russian physiologist I.P. Pavlov, one that belittles his importance to Soviet physiology." . . . *[Almost all speakers] had critical things to say about Bernstein's book. (Sirotkina, 1995, p. 30)*

Critical papers appeared in, among others, the *Komsokol'skaia Pravda* and in *Sovetskii Sport* (Sirotkina, 1995; Feigenberg and Latash, 1996). In the general surge of anti-Semitism, the Jew Bernstein was accused of belittling Pavlov, of relying too much on foreign authors, of developing a theory devoid of practical importance—in short, of "hack work in the guise of science" (*Pravda*, quoted by Sirotkina, 1995, p. 31; cf. Feigenberg and Latash, 1996).

This was a case of very careful timing, probably started from within the Central Committee (Pickenhain, 1998a). A scientific commission had been installed by the Academy of Sciences and the Academy of Medical Sciences, with Aryapetyans as secretary, Bykov as chairman, and Ivanov-Smolensky to second the scientific attacks (Pickenhain, 1998a). They decided beforehand who would suffer, while Bernstein was a special case because of his Stalin Prize. In view of Bernstein's prestige, the main attack on him preceded the 1950 joint meeting, and Bernstein did not even have to be present to know his fate.

In fact, Bernstein was mentioned only twice in the official papers: once by Biriukov because he had "ignored the Pavlov doctrine" and once by Asratyan because "Bernstein knows neither the letter nor the spirit of Pavlov's teachings" (both quotes from Sirotkina, 1995, p. 31). In the discussion, Smirnov revealed that Bernstein now was really past tense:

Prof. Bernstein formulates fantastic hypotheses on the nature of movement coordination and attempts to reject Pavlov's theory with a priori *arguments. He comes to the wrong conclusion that every movement that has become automatic is executed by itself without participation of the highest parts of the central nervous system. Although Bernstein's data are interesting, he has made them unusable with his completely wrong arguments which are factually as well as methodologically incorrect. Now we have the great challenge to elaborate his kymocyclographic results on the basis of Pavlov's theory. (Smirnov, 1950, quoted from the 1954 edition; my translation from the German)*

Even Luria had to denounce Bernstein. Pickenhain (1998a, p. 400) cites Luria from a 1951 meeting of neurologists and psychiatrists:

In my work I failed to take my starting point in Pavlov's theory of the motor analyzer, basing myself instead on the wrong physiological conceptions of P.K. Anokhin and N.A. Bernstein. . . . So, it was impossible for me to approach motor impairments scientifically. The criticism presented is, in its full extent, also valid for my own work, in particular my books on traumatic aphasia and on post-traumatic regeneration of cerebral function. These books are not founded on Pavlov's ideas, and their basic assumptions are thus wrong. (my translation)

To which Pickenhain adds: "This is the language one has to use in a dictatorial system if one is to survive. This is not a matter of guilt of the scientist who utters these lies and self-accusations in order to allow himself to continue to work; it is only the dictatorial system that forces him to do so, that is guilty." (ibid.).

In the aftermath of the Pavlov affair, Bernstein and his wife became addicted to morphine (Feigenberg and Latash, 1996). Still, all was not lost, as in fact Bernstein had already developed a new line of attack on Pavlovianism. In a manuscript written in 1945-1946 but not published at the time (Bernstein, 1996), Bernstein emphasized that in Pavlov's view the organism is passive and can only reach equilibrium with its environment, whereas in reality it is active, all the time changing its environment. It was this "physiology of activity" (cf. Meijer and Bongaardt, 1998; Feigenberg and Meijer, 1999) that would carry him through the rest of his life and would be such an inspiration for his inner circle.

In 1957, riding the crest of cybernetics, Bernstein was back attending conferences and stated, not without revenge:

The period of struggle towards the recognition of the biological importance, the reality and the generality of the principle of cyclical regulation of life processes is now behind us. . . . The debate, in these initial stages, was conducted sharply, but now seems to be over. (Bernstein, 1967b, p. 115; cf. Bongaardt and Meijer, 2000)

Of course, the above is an overstatement, but there was some reason for optimism. Bernstein was entering a relationship with the famous mathematicians Gel'fand and Tsetlin (cf. Bongaardt and Meijer, 2000), and a group of friends had developed around him, each with their own emphasis but still working in the framework of Bernstein's theory—such as Iosif M. Feigenberg, who emphasized probabilistic prognostication, and Lev P. Latash, who stressed the importance of the biological relevance of signals (cf. Graham, 1987).

Thus, during the 1962 meeting (see introduction), Bernstein was not alone. There was critique, but the group could handle it. The most beautiful example of "rubbing it in" is one I found in a recently published translation of a paper by Bassin, Bernstein, and Latash (cf. Latash, Latash, and Meijer, 1999, 2000), written just after the 1962 meeting. The paper deals with a theory of the structure-function relationship in the brain in which the functional organization of the brain changes completely whenever the organism switches to another function. It is hard to get farther away from Pavlov's original views. The authors state:

Each of the points mentioned has highlighted in its own way that the organism, in adapting to the environment, always acts as a fundamentally active structure, as a system whose dynamics are never defined exclusively by factors of the environment. By acknowledging these characteristics, we agree with the dialectic view according to which an external factor becomes a physiological factor only after it has been processed by the brain in the framework of the organism's needs. It becomes clear why, during the last 10 years, neurophysiology has paid so much attention to the organism's needs, why whole brain structures related to motivation and sanctioning were identified, . . . and why it is hard to find any notion in recent published work as deeply elaborated both experimentally and theoretically as the classical Pavlovian notion of reinforcement. (Bassin, Bernstein and Latash, 1999, quoted from Latash, Latash, and Meijer, 2000, pp. 334-335)

The paper appeared in the year of Bernstein's death, 1966, and I cannot help but see him smiling from the grave: Pavlov's epigones are bad communists (bad dialecticians), bad physiologists (running behind), and . . . bad Pavlovians (missing the importance of reinforcement).

This time, there was no remaining silent because the adversary had died. In 1966, Asratyan wrote that Bernstein had "decided to exploit the force of the high-flown word and the effect of outlandish terminology," while his "word tricks and ultramodern terminology are designed to give the author's views the appearance of originality, novelty, and progressiveness" (Asratyan, 1966, quoted from Sirotkina, 1995, pp. 33-34). Asratyan had been away in the Far East when Bernstein died (Feigenberg, personal communication, 1999), and one of his coworkers had allowed the funeral to take place in Asratyan's Institute of Higher Nervous Activity, where "Pavlov's portrait was staring down at Bernstein's coffin" (Bongaardt and Meijer, 2000, p. 69). There appears to have been justice in the Soviet Union after all.

Conclusion

In fact, I have used the notion "Pavlov affair" with two different meanings: as the events surrounding the 1950 Pavlov session and as the whole development of Bernstein's struggle with Pavlov and the Pavlovians. What was it all about?

Personally, I think it was about a woman, but that speculation is at risk of remaining unfalsifiable forever.

Clearly, the Pavlov affair had to do with power: the government's power to interfere with science, Pavlov's power from within his "towers of silence," Bernstein's craving for recognition, unmitigated power for the Pavlovians in their coup of 1948-1950, and finally, a kind of "combative cooperation" (Todes, 1995) where sides may have needed each other, were it only to fight in relative safety (true, with all sincere emotions that were involved).

From the point of view of philosophy of science, the Pavlov affair is absolutely fascinating. To the best of my knowledge, it is the only case where undue external factors (the government, forcing science into dialectical materialism in 1928-1930, but certainly not the 1950 meeting) helped science to improve. Lakatos (e.g., 1980) would not have liked that, but then Stalinism was a two-edged sword.

To the Bernsteinians, the Pavlov affair was about content. It was about the nature of life, the organization of the brain, the coordination of movement. I couldn't agree more.

Somewhere in the 16th century, in the circles around Charles V (Meijer, 2001) and the University of Salamanca (Bandrés and Llavona, 1992), we lost the ability to understand movement. By starting to view nature as a system of clockworks, and the body as an automaton, we forced biological principles of organization to become an enigma. Even Newton himself (1952) understood that a mechanical system cannot control itself. Of course, one could always escape to dualism, but in the end dualism cannot make sense. It was Bernstein who revealed that a purely mechanical view can never capture the organization of living movement. In that sense, I am a Bernsteinian, and I know that the struggle is not over yet.

What do you think of that? What does it mean? Why, it's simply preposterous!
(Pavlov, as cited earlier)

Acknowledgments

Meeting with friends from the larger Bernstein circle was one of the best things that happened in my life, both professionally and personally. I thank them all. Thanks are due also to Peter J. Beek, G. Sander de Wolf, Stephan Praet, and Piet van Wieringen (Amsterdam, The Netherlands), Wu WenHua (Quanzhou, China), and Mark L. Latash (State College, Pennsylvania) for their comments on an earlier version of this chapter. In particular, I want to thank Lothar and Margot Pickenhain for their wonderful hospitality and the many enlightening discussions we had. Of course, none of the aforementioned carries any responsibility whatsoever for the views presented in this chapter.

References

Adrian, E.D. (1936) The late Professor Pavlov. *British Medical Journal,* 1936, 560-561.
Akademija Nauk S.S.S.R., and Akademija Meditsinskikh Nauk S.S.S.R. (1950) *Nauchnaja Sessija Posvyashchennaja Problemam Fiziologicheskogo Uchenija Akademika I.P. Pavlova* [Scientific session dedicated to the problems of the physiological theory of academician I.P. Pavlov]. Moscow: Izdatel'stvo Akademii Nauk S.S.S.R. (Stenographical account, 28 June–4 July.)
Aksyonov, V. (1993-1994). *Moskovskaja saga: Trilogija* [Moscow saga: A trilogy]. Moscow: Tekst. (Appeared in English translation under the title: *Generations of Winter.*)
Andronov, A.A., and Chaikin, C.E. (1949) *Theory of oscillations.* Princeton, NJ: Princeton University Press. (Original work published in Russian in 1937.)
Ash, M. (1995) *Gestalt psychology in German culture, 1890-1967: Holism and the quest for objectivity.* Cambridge, MA: University Press.
Asratyan, E.A. (1953) *I.P. Pavlov: His life and work.* Moscow: Foreign Languages Publishing House. (Original work published in Russian in 1949.)
Babkin, B.P. (1974) *Pavlov: A biography.* Chicago: University of Chicago Press. (Original work published in 1949.)

Bandrés, J., and Llavona, R. (1992) Minds and machines in Renaissance Spain: Gómez Perriera's theory of animal behavior. *J Hist Behav Sci* 28: 158-168.

Barcroft, J. (1936) Obituary: Prof. I.P. Pavlov, For. Mem. R.S. *Nature* 137: 483-484.

Beek, P.J., and Meijer, O.G. (1999) Spinal anticipation and cortical correction: Coordination of movements (1930). *Motor Control* 3: 2-8.

Bernstein, N.A. (1927a) Kymozyclographion, ein neuer Apparat für Bewegungsstudium [Kymocyclography, a new apparatus for the study of movement]. *Pflügers Arch für die gesammte Physiologie des Menschen und der Tiere* 217: 782-792.

Bernstein, N.A. (1927b) Analyse aperiodischer trigonometrischer Reihen [The analysis of aperiodic trigonometric series]. *Z angew Math Mechan* 7: 476-485.

Bernstein, N.A. (1936) Sovremennye iskanija v fiziologii nervnogo protsessa [Contemporary research in the physiology of nervous processes]. (Corrected proofs for a 448-page book, preserved by I.M. Feigenberg.)

Bernstein, N.A. (1947) *O Postroenii Dviženii* [On the construction of movements]. Moscow: Medgiz.

Bernstein, N.A. (1967a) Some emergent problems of the regulation of motor acts. In Bernstein, N.A. (Ed.), *The Co-ordination and regulation of movements,* pp. 15-59. Oxford: Pergamon Press. (Original work published in Russian in 1957.)

Bernstein, N.A. (1967b) The problem of the interrelation of co-ordination with localization. In Bernstein, N.A. (Ed.), *The co-ordination and regulation of movements,* pp. 114-142. Oxford: Pergamon Press. (Original work published in Russian in 1935.)

Bernstein, N.A. (1988a) Auf den Wegen zu einer Biologie der Aktivität [On the roads towards a biology of activity]. In Pickenhain, L., and Schnabel, G. (Eds. and Trans.), *Bewegungsphysiologie von N.A. Bernstein* (2nd ed.), pp. 233-247. Leipzig: Johann Ambrosius Barth. (Original work published in 1965.)

Bernstein, N.A. (1988b) Das Problem der Wechselbeziehungen zwischen Koordination und Lokalisation [The problem of the interrelationships between coordination and localization]. In Pickenhain, L., and Schnabel, G. (Eds. and Trans.), *Bewegungsphysiologie von N.A. Bernstein* (2nd ed.), pp. 67-98. Leipzig: Johann Ambrosius Barth. (Original work published in Russian in 1935.)

Bernstein, N.A. (1996) On dexterity and its development. In Latash, M.L., and Turvey, M.T. (Eds.), *Dexterity and its development,* pp. 3-244. Mahwah, NJ: Lawrence Erlbaum. (Original Russian manuscript written in 1945-1946 and published in 1991.)

Bernstein, N.A., and Popova, T.S. (1929) Untersuchung über die Biodynamik des Klavieranschlags [Study of the biodynamics of piano playing]. *Arbeitsphysiol* 1: 396-432.

Bongaardt, R. (1996) *Shifting focus: The Bernstein tradition in movement science* (PhD thesis). Amsterdam: Rob Bongaardt.

Bongaardt, R., and Meijer, O.G. (2000) Bernstein's theory of movement behavior: Historical development and contemporary relevance. *J Motor Behav* 32: 57-71.

Braune, W., and Fischer, O. (1895-1904) Der Gang des Menschen [Human gait]. *Abhandlungen der Könichlich Sächsischen Gesellschaft der Wissenschaften* 21, 25, 26, and 28 (6 volumes).

Darlington, C.D. (1949) The retreat from science in Soviet Russia. In Zirkle, C. (Ed.), *Death of a Science in Russia,* pp. 67-80. Philadelphia, PA: University of Pennsylvania Press.

Editorial (1936) Ivan Pavlov. *British Medical Journal,* 1936, 507-508.

Feigenberg, I.M. (1988) Chronologisches Verzeichnis aller Publikationen N.A. Bersteins [Chronological list of all Bernstein's publications]. In Pickenhain, L., and Schnabel, G. (Eds.), *Bewegungsphysiologie von N.A. Bernstein,* pp. 255-263. Leipzig: Barth.

Feigenberg, I.M., and Latash, L.P. (1996) N.A. Bernstein: The reformer of neuroscience. In Latash, M.L., and Turvey, M.T. (Eds.), *Dexterity and its development,* pp. 247-275. Mahwah, NJ: Lawrence Erlbaum.

Feigenberg, I.M., and Meijer, O.G. (1999) The active search for information: From reflexes to the model of the future (1966). *Motor Control* 3: 225-236.

Fel'dman, A.G., and Meijer, O.G. (1999) Discovering the right questions in motor control: Movements (1929). *Motor Control* 3: 105-134.

Frolov, I.P. (1937) *Pavlov and his school: The theory of conditioned reflexes.* London: Kegan Paul, Trench, Trubner & Co. (Original work published in Russian in 1937.)

Gel'fand, I.M., Gurfinkel, V.S., Fomin, S.V., and Tsetlin, M.L. (1971) In memory of N.A. Bernstein. In Gel'fand, I.M., Gurfinkel, V.S., Fomin, S.V., and Tsetlin, M.L. (Eds.), *Models of the structural-*

functional organization of certain biological systems, pp. xxxiii-xxxv. Cambridge, MA: MIT Press. (Original work published in Russian in 1966.)

Graham, L.R. (1987) *Science, philosophy and human behavior in the Soviet Union.* New York: Columbia University Press.

Gray, J.A. (1979) *Pavlov.* Brighton: Harvester Press.

Grigorian, N.A. (1974) Pavlov, Ivan Petrovich. In Gillespie, C.G. (Ed.), *Dictionary of scientific biography, volume 10,* pp. 431-436. New York: Scribner's.

Gurfinkel, V.S. (1988) Nikolai Alexandrowitsch Bernstein: Abriß seines wissenschaftlichen Wirkens [Nikolai Aleksandrovich Bernstein: A summary view of his scientific influence]. In Pickenhain, L., and Schnabel, G. (Eds.), *Bewegungsphysiologie von N.A. Bernstein,* pp. 11-14. Leipzig: Johann Ambrosius Barth.

Gurfinkel, V.S., and Cordo, P.J. (1998) The scientific legacy of Nikolai Bernstein. In Latash, M.L. (Ed.), *Progress in motor control, volume 1: Bernstein's traditions in movement science,* pp. 1-19. Champaign, IL: Human Kinetics.

Haken, H. (1977) *Synergetics: An introduction.* Heidelberg: Springer.

Hill, A.V. (1936) A tribute to Pavlov. *British Medical Journal,* 1936, 508-509.

Huxley, J. (1949) *Soviet genetics and world science: Lysenko and the meaning of heredity.* New York: Schuman.

Joravsky, D. (1970) *The Lysenko affair.* Cambridge, MA: Harvard University Press.

Judson, H.F. (1979) *The eighth day of creation: The makers of the revolution in biology.* New York: Simon and Schuster.

Köhler, W. (1973) *Intelligenz Prüfungen an Menschaffen* [Intelligence testing with *anthropoids*]. Springer: Berlin. (Originally published in 1917.)

Kosmodemyansky, A.A. (1956) *Konstantin Tsiolkovsky: His life and work.* Moscow: Foreign Languages Publishing House.

Kozulin, A. (1984) *Psychology in utopia: Toward a social history of Soviet psychology.* Cambridge, MA: MIT Press.

Lakatos, I. (1980) *The methodology of scientific research programmes, philosophical papers, volume 1.* Cambridge: Cambridge University Press. (Original posthumous publication, 1978.)

Latash, L.P., Latash, M.L., and Meijer, O.G. (1999) Thirty years later: On the problem of the relation between structure and function in the brain from a contemporary viewpoint (1966): Part 1. *Motor Control* 3: 329-345.

Latash, L.P., Latash, M.L., and Meijer, O.G. (2000) Thirty years later: On the problem of the relation between structure and function in the brain from a contemporary viewpoint (1966): Part 2. *Motor Control* 4: 125-149.

Lekhtman, Y.B. (1963) [Discussion of Bernstein's paper.] In Akademija Nauk S.S.S.R., Institut Filosofii, *Filosofskije Voprosy Fiziologii Vysshei Nervnoi Dejatel'nosti i Psikhologii,* pp. 552-559. Moscow: Izdatel'stvo Akademii Nauk S.S.S.R.

Luria, A.R. (1987) Bernstein, Nicholas. In Gregory, R.L. (Ed.), *The Oxford compendium to the mind,* pp. 805-806. Oxford: Oxford University Press.

Lysenko, T.D. (1954) *Agrobiology: Essays on problems of genetics, plant breeding and seed growing.* Moscow: Foreign Languages Publishing House.

Mecacci, L. (1979) *Brain and history: The relationship between neurophysiology and psychology in Soviet research.* New York: Brunner/Mazel. (Original work published in Italian in 1977.)

Meijer, O.G. (1985) Hugo de Vries no Mendelian? *Ann Sci* 42: 189-232.

Meijer, O.G. (2001) Making things happen: An introduction to the history of movement science. In Latash, M.L., and Zatsiorsky, V.M. (Eds.), *Classics in movement science,* pp. 1-57. Champaign, IL: Human Kinetics.

Meijer, O.G., and Bongaardt, R. (1998) Bernstein's last paper: The immediate task of neurophysiology in the light of the modern theory of biological activity. *Motor Control* 2: 2-9.

Meijer, O.G., and Feigenberg, I.M. (2000). Bernstein's failure to join the space race: His commentary on Tsiolkovskii's "Mechanics in biology" (1964). *Motor Control* 4: 262-272.

Meijer, O.G., and Wagenaar, R.C. (1998) Bernstein's rejection of Braune and Fischer: Studies on the physiology and pathology of movements (1936). *Motor Control* 2: 95-100.

Newton, I. (1952) *Opticks.* New York: Dover. (Original work published 1704.)

Pavlov, I.P. (1955a) [Criticism of the *Gestalt* psychology.] In *I.P. Pavlov: Selected works,* pp. 569-576. Moscow: Foreign Language Publishing House. (Original statement in Russian, 1934b.)

Pavlov, I.P. (1955b) [Criticism of Sherrington's idealistic concepts.] In *I.P. Pavlov: Selected works,* pp. 563-569. Moscow: Foreign Language Publishing House. (Original statement in Russian, 1934a.)

Pavlov, I.P. (1955c) Dynamic stereotypy of the higher part of the brain. In *I.P. Pavlov: Selected works,* pp. 448-453. Moscow: Foreign Language Publishing House. (Original statement 1932b, on the occasion of the Tenth International Congress of Psychologists in Copenhagen.)

Pavlov, I.P. (1955d) Reply of a physiologist to psychologists. In *I.P. Pavlov: Selected works,* pp. 409-447. Moscow: Foreign Language Publishing House. (Original publication in *Psychol Rev* 39, 1932a.)

Pavlov, I.P. (1955e) [Speech at the opening of the Fifteenth International Physiological Congress.] In *I.P. Pavlov: Selected works,* pp. 56-58. Moscow: Foreign Language Publishing House. (Original speech given in Russian in 1935.)

Pavlov, I.P. (1955f) The nature of intelligence in *anthropoids* and the erroneous interpretation of Koehler. In *I.P. Pavlov: Selected works,* pp. 558-562. Moscow: Foreign Language Publishing House. (Original statement in Russian, 1934c.)

Pflüger, E.F.W. (1853) *Die sensorischen Funktionen des Rückenmarks der Wirbelthiere nebst einer neuen Lehre über die Leitungsgezetze der Reflexionen* [The sensory functions of the spinal cord of vertebrates plus a new theory of the conduction laws of reflexes]. Bonn: Max Cohen & Sohn.

Pickenhain, L. (1998a) Das Schicksal der Pawlowschen Ideeen in der UdSSR [The fate of Pavlov's ideas in the Soviet Union]. In Pickenhain, L. (Ed.), *I.P. Pavlov: Gesammelte Werke über die Physiologie und Pathologie der höheren Nerventätigkeit,* pp. 373-406. Würzburg: Ergon.

Pickenhain, L. (1998b) Vorwort [Preface]. In Pickenhain, L. (Ed.), *I.P. Pavlov: Gesammelte Werke über die Physiologie und Pathologie der höheren Nerventätigkeit,* pp. 7-9. Würzburg: Ergon.

Pickenhain, L. (1998c) Einleitung [Introduction]. In Pickenhain, L. (Ed.), *I.P. Pavlov: Gesammelte Werke über die Physiologie und Pathologie der höheren Nerventätigkeit,* pp. 11-21. Würzburg: Ergon.

Pickenhain, L. (Ed.) (1998d) *I.P. Pavlov: Gesammelte Werke über die Physiologie und Pathologie der höheren Nerventätigkeit* [I.P. Pavlov: Collected works on the physiology and pathology of higher nervous activity]. Würzburg: Ergon.

Pickenhain, L., and Schnabel, G. (1988) Einführung [Introduction]. In Pickenhain, L., and Schnabel, G. (Eds.), *Bewegungsphysiologie von N.A. Bernstein,* pp. 15-19. Leipzig: Johann Ambrosius Barth.

Planck, M. (1910) *Acht Vorlesungen über theoretische Physik* [Eight lectures about theoretical physics]. Leipzig: Hirzel.

Popper, K.R., and Eccles, J.C. (1977) *The self and its brain.* Berlin: Springer.

Powers, J. (1982) *Philosophy and the new physics.* London: Methuen.

Prigogine, I., and Stengers, I. (1984) *Order out of chaos.* New York: Bantam.

Regelmann, J.-P. (1980) *Die Geschichte des Lyssenkoismus* [The history of Lysenkoism]. Frankfurt: Rita Fischer Verlag.

Sherrington, C. (1933) *The brain and its mechanism.* Cambridge: At the University Press.

Sherrington, C. (1952) *The integrative action of the nervous system.* Cambridge: At the University Press. (Original work published in 1906.)

Sherrington, C. (1955) *Man on his nature.* Harmondsworth: Penguin. (Original publication in 1940.)

Sherrington, C. (1979) *Selected writings.* Oxford: Oxford University Press. (First published in 1939, edited by D. Danny-Brown.)

Sirotkina, I.E. (1995) N.A. Bernstein: The years before and after "the Pavlov session." *Russian Studies in History,* Fall 1995: 24-36.

Smirnov, K.M. (1954) [Discussion.] In *Wisisenschaftliche Tagung über die Probleme der physiologischen Lehre I.P. Pavlov's, gemeinsam durchgeführt von der Akademie der Wissenschaften und der Akademie der Medizinischen Wissenschaften der UdSSR in Moskau, 28. Juni bis 4. Juli 1950, 4. Heft* [Scientific Meeting on the problems of the physiological theory of I.P. Pavlov, organized by the Academy of Sciences and the Academy of Medical Sciences of the Soviet Union, June 28-July 4, 1950, in Moscow, Volume 4], pp. 520-524. Berlin: Verlag Kultur und Fortschritt. [German translation of the stenographical notes.]

Solandt, D.Y. (1935) International Physiological Congress: Meeting in the U.S.S.R. *Nature* 136: 571-575.

Sporns, O., and Edelman, G.M., with endnotes provided by Meijer, O.G. (1998). Bernstein's dynamic view of the brain: The current problems of modern neurophysiology (1945). *Motor Control* 2: 283-305.

Starling, E.H. (1925) Ivan Petrovich Pavlov. *Nature* 115: 1-3.

Todes, D.P. (1995) Pavlov and the Bolsheviks. *History and Philosophy of the Life Sciences* 17: 379-418.

Weiss, P.A. (1959) Animal behavior as system reaction: Orientation toward light and gravity in the resting position of butterflies *(Vanessa)*. In Bertalanffy, L. von, and Rappaport, A. (Eds.), *General systems: Yearbook of the Society for General Systems Research,* pp. 1-44. Ann Arbor, MI: Society for General Systems Research. (Original work published in German in 1925.)

Windholz, G. (1983) Pavlov's position toward American behaviorism. *J Hist Behav Sci* 19: 394-407.

Windholz, G. (1990) Pavlov and the Pavlovians in the laboratory. *J Hist Behav Sci* 26: 64-74.

Index

Note: The italicized *f* and *t* following page numbers refer to figures and tables, respectively.

Contributors

Philippe Archambault, Neurological Science Research Centre, University of Montréal; and Research Centre, Rehabilitation Institute of Montréal, Montréal, Quebec, Canada.

Y.I. Arshavsky, Institute for Nonlinear Science, University of California at San Diego, La Jolla, California, U.S.A., and Institute of Information Transmission Problems, Russian Academy of Sciences, Moscow, Russia.

Andrew G. Barto, Department of Computer Science, University of Massachusetts, Amherst, Massachusetts, U.S.A.

Simon Bouisset, Laboratoire de Physiologie du Mouvement, Université Paris-Sud, Orsey, France.

T.J. Burkholder, Department of Applied Physiology, Georgia Institute of Technology, Atlanta, Georgia, U.S.A.

Evangelos A. Christou, Department of Kinesiology and Applied Physiology, University of Colorado at Boulder, Colorado, U.S.A.

Carmen M. Cirstea, Neurological Science Research Centre, University of Montréal; and Research Centre, Rehabilitation Institute of Montréal, Montréal, Quebec, Canada.

T.G. Deliagina, The Nobel Institute of Neurophysiology, Department of Neuroscience, Karolinska Institutet, Stockholm, Sweden.

Roger M. Enoka, Department of Kinesiology and Applied Physiology, University of Colorado at Boulder, Colorado, U.S.A.

Andrew H. Fagg, Department of Computer Science, University of Massachusetts, Amherst, Massachusetts, U.S.A.

R. Gantcheva, Service de Neurologie, CHG d'Aix-en-Provence, France.

Israel M. Gelfand, Rutgers University, New Brunswick, New Jersey, U.S.A.

Apostolos P. Georgopoulos, Brain Sciences Center, Minneapolis Veterans Affairs Medical Center; Departments of Neuroscience, Neurology, and Psychiatry, University of Minnesota Medical School; and Cognitive Sciences Center, University of Minnesota, Minneapolis, Minnesota, U.S.A.

James C. Houk, Department of Physiology, Northwestern University School of Medicine, Chicago, Illinois, U.S.A.

M. Ioffe, Institute of Higher Nervous Activity and Neurophysiology, Russian Academy of Sciences, Moscow, Russia.

Serge Le Bozec, Laboratoire de Physiologie du Mouvement, Université Paris-Sud, Orsey, France.

Mindy F. Levin, School of Rehabilitation, University of Montréal; and Research Centre, Rehabilitation Institute of Montréal, Montréal, Quebec, Canada.

J. Massion, UPR Neurobiologie et Mouvement, Centre National de la Recherche Scientifique, Marseille, France.

Onno G. Meijer, Faculty of Human Movement Sciences, Vrije Universiteit, Amsterdam, The Netherlands.

T.R. Nichols, Department of Physiology, Emory University, Atlanta, Georgia, U.S.A.

G.N. Orlovsky, The Nobel Institute of Neurophysiology, Department of Neuroscience, Karolinska Institutet, Stockholm, Sweden.

Agnès Roby-Brami, Centre Nationale de la Recherche Scientifique, Paris, France.

J.C. Rothwell, Sobell Department, Institute of Neurology, London, United Kingdom.

C. Schmitz, UPR Neurobiologie et Mouvement, Centre National de la Recherche Scientifique, Marseille, France.

Florina Son, Neurological Science Research Centre, University of Montréal; and Research Centre, Rehabilitation Institute of Montréal, Montréal, Quebec, Canada.

Brian L. Tracy, Department of Kinesiology and Applied Physiology, University of Colorado at Boulder, Colorado, U.S.A.

J. Valls-Solé, Unitat d'EMG, Servei de Neurologia, Hospital Clinic, Barcelona, Spain.

F. Viallet, Service de Neurologie, CHG d'Aix-en-Provence, France.

R.J.H. Wilmink, Department of Physiology, Emory University, Atlanta, Georgia, U.S.A.

Marjorie Hines Woollacott, Department of Exercise and Movement Science and Institute of Neuroscience, University of Oregon, Eugene, Oregon, U.S.A.

About the Editor

Mark L. Latash, PhD, is a professor of kinesiology at Penn State University. Since the 1970s, he has worked extensively in normal and disordered motor control. His work has included animal studies, human experiments, modeling, and clinical studies.

The author of *Control of Human Movement* (Human Kinetics, 1993) and *Neurophysiological Basis of Movement* (Human Kinetics, 1998), Latash also translated Bernstein's classic, *On Dexterity and Its Development* (Erlbaum) in 1996. He serves as the editor of the academic journal *Motor Control* and was a coauthor of *Classics in Movement Science* (Human Kinetics, 2001).

Latash earned a master's degree in physics of living systems from the Moscow Physico-Technical Institute in 1976 and a PhD in physiology from Rush University in 1989. He is a member of the Society for Neuroscience and the American Society of Biomechanics.